The F

Energy Cover-up

A Hundred Year Deception

By Ronald E. Hudkins

Copyright Notice

ISBN: 9781795008723

Book Summary

Research appears to show that over the greater span of American history a number of significant breakthroughs in energy generation and propulsion have occurred that have been systematically suppressed. For example; Nikola Tesla discovered Zero Point energy in 1902, but the technology was suppressed by the government. Since then, 5, 000 patents on technology of this kind have been legally confiscated.

In doing this, our government has committed possibly the greatest injustice against humanity, ever. If just Tesla's discovery of zero point energy had become common knowledge back in the 1900s, and the technology developed and put into widespread use, we could have had virtually free, unlimited, non-polluting energy for nearly 120 years.

We have had the technological ability to replace fossil fuel, internal combustion and nuclear power generating systems with advanced non-polluting electromagnetic and electro-gravity systems. The open literature is replete with well-documented technologies that have surfaced, only to later be illegally seized or suppressed through systematic abuses of the national security state, large corporate and financial interests or other shadowy concerns.

The work of Tesla, other scientists, backyard inventors' interviews and whistleblowers is a common theme in this book. The overall compilation of publications provide a powerful driving force for us to put pressure on our elected officials and demand that they support the development and perforation into the marketplace of clean energy devices and infrastructure – with the same vigor with which they approached the Manhattan project or the Apollo Project. Help spread our technological knowledge awareness.

Table of Contents

Car Achieves Almost 10,000 Miles Per Gallon

Was Edison Adversary Father Of 'Star Wars'?

Grid parity: Why electric utilities should struggle to sleep at night

Miracle material graphene can distil booze, says study

Abiotic Oil: A Theory Worth Exploring

Blacklight Power bolsters its impossible claims of a new renewable energy source

Puzzled Researchers Vet BlackLight's Physics-Defying Hydrogen Power

Loremo: The 'Low Resistance Mobile'

The New Dawn of Solar

Bribes Offered to Scientists

Kids Build Soybean-Fueled Car

Eco-car more efficient than light bulb

Refiners Maintain a Firm but Legal Grip on Supplies

Enron: The Smartest Guys in the Room

Toyota smashes fuel economy record

Listen, Detroit: You'll Get a Charge Out of This

Method and Apparatus for Tunneling by Melting

Lockheed Martin Receives Patent For 'World Changing' Fusion Reactor

The US government keeps spectacularly underestimating solar energy installation

Tiny turbine that fits on your DESK runs on carbon dioxide - and it can produce enough energy to power a small town

13-Year-Old Creates Energy Harvesting Device

The sports car that runs on saltwater

Tesla's Model S receives top rating from Consumer Reports

Toyota Rav4 EV review: electrifying

Are we wildly underestimating solar and wind power?

The extra dollars you're paying at the pump are going to Wall Street speculators

Method for Enhancement of Surface Plasmon Polaritons to Initiate & Sustain LENR (Cold Fusion)

Solar Energy: The Quest for Cheap

Man makes drinking water out of thin air

Who Killed the Electric Car

Safe nuclear does exist, and China is leading the way with thorium

Fusion's Ups and Downs

The Bloom Box: An Energy Breakthrough?

Solar Energy All Night Long

Cars that make hybrids look like gas guzzlers

Coming in out of the cold: Cold fusion, for real

Fans of GM Electric Car Fight the Crusher

Air Force pursuing antimatter weapons

This breakthrough in a type of photosynthesis could provide the world with unlimited energy

We're drowning in trash. These Dutch scientists have a solution.

Costa Rica to ban fossil fuels and become world's first decarbonized society

Power Prices Go Negative in Germany, a Positive for Energy Users

This Incredible Power Plant Produces Nothing But Electricity And Stone

GM is selling a $5,000 electric car in China

Scientists flip energy equation with solar leaf that converts CO2 into fuel

Long-lasting flow battery could run for more than a decade with minimum upkeep

NASA approves 'impossible' space engine design that apparently violates the laws of physics and could revolutionize space travel

Power shifts as off-grid options spread worldwide

Introduction

The World of Free Energy

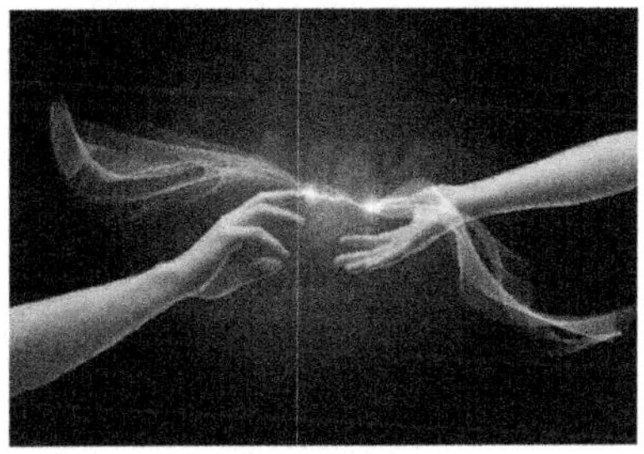

In the late 1880's, trade journals in the electrical sciences were predicting free electricity and free energy in the near future. Incredible discoveries about the nature of electricity were becoming common place. Nikola Tesla was demonstrating "wireless lighting" and other wonders associated with high frequency currents. There was an excitement about the future like never before.

Within 20 years, there would be automobiles, airplanes, movies, recorded music, telephones, radio, and practical cameras. The Victorian Age was giving way to something totally new. For the first time in history, common people were encouraged to envision a utopian future filled with abundant modern transportation and communication, as well as jobs, housing and food for everyone. Disease would be conquered, and so would poverty. Life was getting better, and this time, everyone was going to get a piece of the pie.

So, what happened? In the midst of this technological explosion, where did the energy breakthroughs go? Was all of this excitement about free energy, which happened just before the beginning of the last century, just wishful thinking that "real science" eventually disproved?

Current State of Technology

Actually, the answer to that question is no. In fact, the opposite is true. Spectacular energy technologies were developed right along with the other breakthroughs. Since that time, multiple methods for producing vast amounts of energy at extremely low cost have been developed. None of these technologies have made it to the consumer market as an article of commerce, however. Exactly why this is true will be discussed shortly.

But first, I would like to describe to you a short list of free energy technologies that I am currently aware of, and that are proven beyond all reasonable doubt. The common feature connecting all of these discoveries, is that they use a small amount of one form of energy to control or release a large amount of a different kind of energy. Many of them tap the underlying anther field in some way; a source of energy conveniently ignored by modern science.

1)	**Radiant Energy.** Nikola Tesla's magnifying transmitter, T. Henry Moray's radiant energy device, Edwin Gray's EMA motor, and Paul Baumann's Testatika machine all run on radiant energy. This natural energy form can be gathered directly from the environment (mistakenly called "static" electricity) or extracted from ordinary electricity by the method called fractionation. Radiant energy can perform the same wonders as ordinary electricity, at less than 1% of the cost. It does not behave exactly like electricity, however, which has contributed to the scientific community's misunderstanding of it. The Methernitha Community in Switzerland currently has 5 or 6 working models of fuelless, self-running devices that tap this energy.
2)	**Permanent Magnets.** Dr. Robert Adams (NZ) has developed astounding designs of electric motors, generators and heaters that run on permanent magnets. One such device draws 100 watts of electricity from the source, generates 100 watts to recharge the source, and produces over 140 BTU's of heat in two minutes! Dr. Tom Bearden(USA) has two working models of a permanent magnet powered electrical transformer. It uses a 6-watt electrical input to control the path of a magnetic field coming out of a permanent magnet. By channeling the magnetic field, first to one output coil and then a second output coil, and by doing this repeatedly and rapidly in a "ping-pong" fashion, the device can produce a 96-watt electrical output with no moving parts. Bearden calls his device a Motionless Electromagnetic Generator, or MEG. Jean-Louis Naudin has duplicated Bearden's device in France. The principles for this type of device were first disclosed by Frank Richardson (USA) in 1978. Troy Reed (USA) has working models of a special magnetized fan that heats up as it spins. It takes exactly the same amount of energy to spin the fan whether it is generating heat or not. Beyond these developments, multiple inventors have identified working mechanisms that produce motor torque from permanent magnets alone.
3)	**Mechanical Heaters.** There are two classes of machines that transform a small amount of mechanical energy into a large amount of heat. The best of these purely mechanical designs are the rotating cylinder systems designed by Frenette (USA) and Perkins (USA). In these machines, one cylinder is rotated within another cylinder with about an eighth of an inch of clearance between them. The space between the cylinders is filled with a liquid such as water or oil, and

	it is this "working fluid" that heats up as the inner cylinder spins. Another method uses magnets mounted on a wheel to produce large eddy currents in a plate of aluminum, causing the aluminum to heat up rapidly. These magnetic heaters have been demonstrated by Muller (Canada), Adams (NZ) and Reed (USA). All of these systems can produce ten times more heat than standard methods using the same energy input.
4)	**Super-Efficient Electrolysis**. Water can be broken into hydrogen and oxygen using electricity. Standard chemistry books claim that this process requires more energy than can be recovered when the gases are recombined. This is true only under the worst case scenario. When water is hit with its own molecular resonant frequency, using a system developed by Stan Meyers (USA) and again recently by Xogen Power, Inc., it collapses into hydrogen and oxygen gas with very little electrical input. Also, using different electrolytes (additives that make the water conduct electricity better) changes the efficiency of the process dramatically. It is also known that certain geometric structures and surface textures work better than others do. The implication is that unlimited amounts of hydrogen fuel can be made to drive engines (like in your car) for the cost of water. Even more amazing is the fact that a special metal alloy was patented by Freedman (USA) in 1957 that spontaneously breaks water into hydrogen and oxygen with no outside electrical input and without causing any chemical changes in the metal itself. This means that this special metal alloy can make hydrogen from water for free, forever.
5)	**Implosion/Vortex**. All major industrial engines use the release of heat to cause expansion and pressure to produce work, like in your car engine. Nature uses the opposite process of cooling to cause suction and vacuum to produce work, like in a tornado. Viktor Schauberger (Austria) was the first to build working models of implosion engines in the 1930's and 1940's. Since that time, Callum Coats has published extensively on Schauberger's work in his book *Living Energies* and subsequently, a number of researchers have built working models of implosion turbine engines. These are fuelless engines that produce mechanical work from energy accessed from a vacuum. There are also much simpler designs that use vortex motions to tap a combination of gravity and centrifugal force to produce a continuous motion in fluids.

6)	**Cold Fusion**. In March 1989, two chemists from the University of Utah (USA) announced that they had produced atomic fusion reactions in a simple tabletop device. The claims were "debunked" within six months and the public lost interest. Nevertheless, cold fusion is very real. Not only has excess heat production been repeatedly documented, but also low energy atomic element transmutation has been catalogued, involving dozens of different reactions! This technology definitely can produce low cost energy and scores of other important industrial processes.
7)	**Solar Assisted Heat Pumps**. The refrigerator in your kitchen is the only free energy machine you currently own. It's an electrically operated heat pump. It uses one amount of energy (electricity) to move three amounts of energy (heat). This gives it a co-efficient of performance (COP) of about 3. Your refrigerator uses one amount of electricity to pump three amounts of heat from the inside of the refrigerator to the outside of the refrigerator. This is its typical use, but it is the worst possible way to use the technology. Here's why. A heat pump pumps heat from the source of heat to the "sink" or place that absorbs the heat. The source of heat should obviously be hot and the sink for heat should obviously be cold for this process to work the best. In your refrigerator, it's exactly the opposite. The source of heat is inside the box, which is cold, and the sink for heat is the room temperature air of your kitchen, which is warmer than the source. This is why the COP remains low for your kitchen refrigerator. But this is not true for all heat pumps. COP's of 8 to 10 are easily attained with solar assisted heat pumps. In such a device, a heat pump draws heat from a solar collector and dumps the heat into a large underground absorber, which remains at 55° F, and mechanical energy is extracted in the transfer. This process is equivalent to a steam engine that extracts mechanical energy between the boiler and the condenser, except that it uses a fluid that boils at a much lower temperature than water. One such system that was tested in the 1970's produced 350 hp, measured on a Dynamometer, in a specially designed engine from just 100-sq. ft. of solar collector. (This is not the system promoted by Dennis Lee.) The amount of energy it took to run the compressor (input) was less than 20 hp, so this system produced more than 17 times more energy than it took to keep it going! It could power a small neighborhood from the roof of a hot tub gazebo, using exactly the same technology that keeps the food cold in your kitchen. Currently, there is an industrial scale heat pump

	system just north of Kona, Hawaii that generates electricity from temperature differences in ocean water.

There are dozens of other systems that I have not mentioned, many of them are as viable and well tested as the ones I have just recounted. But this short list is sufficient to make my point: free energy technology is here, now. It offers the world pollution-free, energy abundance for everyone, everywhere.

It is now possible to stop the production of greenhouse gases and shut down all of the nuclear power plants. We can now desalinate unlimited amounts of seawater at an affordable price, and bring adequate fresh water to even the most remote habitats. Transportation costs and the production costs for just about everything can drop dramatically. Food can even be grown in heated greenhouses in the winter, anywhere. All of these wonderful benefits that can make life on this planet so much easier and better for everyone have been postponed for decades. Why? Whose purposes are served by this postponement?

Four Invisible Forces

There are four gigantic forces that have worked together to create this situation. To say that there is and has been a conspiracy to suppress this technology only leads to a superficial understanding of the world, and it places the blame for this completely outside of ourselves. Our willingness to remain ignorant and actionless in the face of this situation has always been interpreted by two of these forces as implied consent. So, besides a non-demanding public, what are the other forces that are impeding the availability of free energy technology?

In the United States, and in most other countries around the world, there is a money monopoly in place. I am free to earn as much money as I want, but I will only be paid in Federal Reserve Notes. There is nothing I can do to be paid in Gold Certificates, or some other form of money. This money monopoly is solely in the hands of a small number of private stock banks, and these banks are owned by the wealthiest families in the world. Their plan is to eventually control 100% of all of the capital resources of the world, and thereby control everyone's life through the availability (or non-availability) of all goods and services.

An independent source of wealth (free energy device) in the hands of each and every person in the world, ruins the plans of the wealthiest families for world domination, permanently. Why this is true is easy to see. Currently, a nation's economy can be either slowed down or sped up by the raising or lowering of interest rates. But if an independent source of capital (energy) were present in the economy, and any business or person could raise more capital without borrowing it from a bank, this centralized throttling action on interest rates would simply not have the same effect.

Free energy technology changes the value of money. The wealthiest families and the issuers of credit do not want any competition. It's that simple. They want to maintain their current monopoly control of the money supply. For them, free energy technology is not just something to suppress, it must be permanently forbidden!

So, the wealthiest families and their central banking institutions are **the first force** operating to postpone the public availability of free energy technology. Their motivations are the imagined divine right to rule, greed, and their insatiable need to control almost everything except themselves. The weapons they have used to enforce this postponement include intimidation, "expert" debunkers, buying and shelving of technology, murder and attempted murder of the inventors, character assassination, arson, and a wide variety of financial incentives and disincentives to manipulate possible supporters. They have also promoted the general acceptance of a scientific theory that states that free energy is impossible (laws of thermodynamics).

The second force operating to postpone the public availability of free energy technology is national governments. The problem here is not so much related to competition in the printing of currency, but in the maintenance of national security. The fact is, the world out there is a jungle, and humans can be counted upon to be very cruel, dishonest, and sneaky. It is government's job to provide for the common defense. For this, police powers are delegated by the executive branch of government to enforce "the rule of law." Most of us who consent to the rule of law do so because we believe it is the right thing to do, for our own benefit. There are always a few individuals, however, that believe that their own benefit is best served by behavior that does not voluntarily conform to the generally agreed upon social order. These people choose to operate outside of the rule of law and are considered outlaws, criminals, subversives, traitors, revolutionaries, or terrorists.

Most national governments have discovered, by trial and error, that the only foreign policy that really works, over time, is a policy called "tit for tat." What this means to you and me is, that governments treat each other the way they are being treated. There is a constant jockeying for position and influence in world affairs, and the strongest party wins! In economics, it's the Golden Rule, which states: "The one with the gold makes the rules."

So it is with politics also, but its appearance is more Darwinian. It's simply survival of the fittest. In politics, however, the fittest has come to mean the strongest party who is also willing to fight the dirtiest. Absolutely every means available is used to maintain an advantage over the adversary, and everyone else is the adversary regardless of whether they are considered friend or foe. This includes outrageous psychological posturing, lying, cheating, spying, stealing, assassination of world leaders, proxy wars, alliances and shifting alliances, treaties, foreign aid, and the presence of military forces wherever possible.

Like it or not, this is the psychological and actual arena national governments operate in. No national government will do anything that simply gives an adversary an advantage for free. It's national suicide. An activity by any individual, inside or outside the country, that is interpreted as giving an adversary an edge or advantage will be deemed a threat to "national security."

Free energy technology is a national government's worst nightmare! Openly acknowledged, free energy technology sparks an unlimited arms race by all governments in a final attempt to gain absolute advantage and domination. Think about it. Do you think Japan will not feel intimidated if China gets free energy? Do you think Israel will sit by quietly as Iraq acquires free energy? Do you think India will allow Pakistan to develop free energy? Do you think the USA would not try to stop Osama bin Laden from getting free energy?

Unlimited energy available to the current state of affairs on this planet leads to an inevitable reshuffling of the balance of power. This could become an all-out war to prevent "the other" from having the advantage of unlimited wealth and power. Everybody will want it, and at the same time, want to prevent everyone else from getting it.

So, national governments are the second force operating to postpone the public availability of free energy technology. Their motivations are "self-preservation." This self-preservation operates on three levels. First, by not giving undue advantage to an external enemy. Second, by preventing individualized action capable of effectively challenging official police powers (anarchy) within the country. And third, by preserving income streams derived from taxing energy sources currently in use. Their weapons include the preventing of the issuance of patents based on national security grounds, the legal and illegal harassment of inventors with criminal charges, tax audits, threats, phone taps, arrest, arson, theft of property during shipment, and a host of other intimidations which make the business of building and marketing a free energy machine practically impossible.

The third force operating to postpone the public availability of free energy technology consists of the group of deluded inventors and out right charlatans and con men. On the periphery of the extraordinary scientific breakthroughs that constitute the real free energy technologies, lies a shadow world of unexplained anomalies, marginal inventions and unscrupulous promoters. The first two forces have constantly used the media to promote the worst examples of this group, to distract the public's attention and to discredit the real breakthroughs by associating them with the obvious frauds.

Over the last hundred years, dozens of stories have surfaced about unusual inventions. Some of these ideas have so captivated the public's imagination that a mythology about these systems continues to this day. Names like Keely, Hubbard, Coler, and Henderschott immediately come to mind. There may be real technologies behind these names, but there simply isn't enough technical data available in the public domain to make a determination. These names remain associated with a free energy mythology, however, and are cited by debunkers as examples of fraud.

The idea of free energy taps very deeply into the human subconscious mind. A few inventors with marginal technologies that demonstrate useful anomalies have mistakenly exaggerated the importance of their inventions. Some of these inventors also have mistakenly exaggerated the importance of themselves for having invented it. A combination of "gold fever" and/or a messiah complex appears, wholly distorting any future contribution they may make.

While the research thread they are following may hold great promise, these deluded inventors begin to trade enthusiasm for facts, and the value of the scientific work from that point on suffers greatly. There is a powerful, yet subtle seduction that can warp a personality if they believe that the world rests on their shoulders or that they are the world's savior. Strange things also happen to people when they think they are about to become extremely rich. It takes a tremendous spiritual discipline to remain objective and humble in the presence of a working free energy machine.

Many inventors' psyches become unstable just believing they have a free energy machine. As the quality of the science deteriorates, some inventors also develop a persecution complex that makes them very defensive and unapproachable. This process precludes them from ever really developing a free energy machine, and fuels the fraud mythologies tremendously.

Then there are the outright con men. In the last 15 years, there is one person in the USA who has raised the free energy con to a professional art. He has raised more than $100,000,000, has been barred from doing business in the state of Washington, has been jailed in California, and he's still at it. He always talks about a variation of one of the real free energy systems, sells people on the idea that they will get one of these systems soon, but ultimately sells them only promotional information which gives no real data about the energy system itself. He has mercilessly preyed upon the Christian community and the patriot community in the USA, and is still going strong.

This man's current scam involves signing up hundreds of thousands of people as locations where he will install a free energy machine. In exchange for letting him put the free energy generator in their home, they will get free electricity for life, and his company will sell the excess energy back to the local utility company. After becoming convinced that they will receive free electricity for life, with no out-front expenses, they gladly buy a video that helps draw their friends into the scam as well. Once you understand the power and motivations of the first two forces I have discussed, its obvious that this person's current business plan cannot be implemented. This one person has probably done more harm to the free energy movement in the USA than any other force, by destroying people's trust in the technology.

So, the third force postponing the public availability of free energy technology is delusion and dishonesty within the movement itself. The motivations are self-aggrandizement, greed, want of power over others, and a false sense of self-importance. The weapons used are lying, cheating, the "bait and switch" con, self-delusion and arrogance combined with lousy science.

The fourth force operating to postpone the public availability of free energy technology is all of the rest of us. It may be easy to see how narrow and selfish the motivations of the other forces are, but actually, these motivations are still very much alive in each of us as well. Like the Wealthiest Families, don't we each secretly harbor illusions of false superiority, and the want to control others instead of ourselves? Also, wouldn't you "sell out" if the price were high enough, say, take a million dollars, cash, today? Or like the governments, don't we each want to ensure our own survival? If caught in the middle of a full, burning theater, do you panic and push all of the weaker people out of the way in a mad, scramble for the door? Or like the deluded inventor, don't we trade a comfortable illusion once in a while for an uncomfortable fact? And don't we like to think more of ourselves than others give us credit for? Or don't we still fear the unknown, even if it promises a great reward?

You see, really, all four forces are just different aspects of the same process, operating at different levels in the society. There is really only one force preventing the public availability of free energy technology, and that is the unspiritually motivated behavior of the humans. In the last analysis, free energy technology is an outward manifestation of divine abundance. It is the engine of the economy of an enlightened society, where people voluntarily behave in a respectful and civil manner toward each other, where each member of the society has everything they need, and does not covet what their neighbor has, where war and physical violence has become socially unacceptable behavior and people's differences are at least tolerated, if not enjoyed.

The appearance of free energy technology in the public domain is the dawning of a truly civilized age. It is an epochal event in human history. Nobody can take credit for it. Nobody can get rich on it. Nobody can rule the world with it. It is simply, a gift from God. It forces us all to take responsibility for our own actions and for our own self-disciplined self-restraint when needed. The world as it is currently ordered, cannot have free energy technology without being totally transformed by it into something else. This civilization has reached the pinnacle of its development, because it has birthed the seeds of its own transformation. Unspiritualized humans cannot be trusted with free energy. They will only do what they have always done, which is take merciless advantage of each other, or kill each other and themselves in the process.

If you go back and read Ayn Rand's *Atlas Shrugged* or the *Club of Rome Report*, it becomes obvious that the wealthiest families have understood this for decades. Their plan is to live in the world of free energy, but permanently freeze the rest of us out. But this is not new. Royalty has always considered the general population (us) to be their subjects. What is new, is that you and I can communicate with each other now better than at any time in the past. The Internet offers us, the fourth force, an opportunity to overcome the combined efforts of the other forces preventing free energy technology from spreading.

The Opportunity

What is starting to happen is that inventors are publishing their work, instead of patenting it and keeping it secret. More and more, people are giving away information on these technologies in books, videos and websites. While there is still a great deal of useless information about free energy on the Internet, the availability of good information is rising rapidly. Check out the list of websites and other resources at the end of this article.

It is imperative that you begin to gather all of the information you can on real free energy systems. The reason for this is simple. The first two forces will never allow an inventor or a company to build and sell a free energy machine to you! The only way you will ever get one is if you, or a friend, build it yourself. This is exactly what thousands of people are already quietly starting to do. You may feel wholly inadequate to the task, but start gathering information now. You may be just a link in the chain of events for the benefit of others. Focus on what you can do now, not on how much there still is to be done. Small, private research groups are working out the details as you read this. Many are committed to publishing their results on the Internet.

All of us constitute the fourth force. If we stand up and refuse to remain ignorant and actionless, we can change the course of history. It is the aggregate of our combined action that can make a difference. Only the mass action that represents our consensus can create the world we want. The other three forces will not help us put a fuelless power plant in our basements. They will not help us be free from their manipulations. Nevertheless, free energy technology is here. It is real, and it will change everything about the way we live, work and relate to each other. In the last analysis, free energy technology obsoletes greed and the fear for survival. But like all exercises of spiritual faith, we must first manifest the generosity and trust in our own lives.

The source of free energy is inside of us. It is that excitement of expressing ourselves freely. It is our spiritually guided intuition expressing itself without distraction, intimidation or manipulation. It is our open-heartedness. Ideally, the free energy technologies underpin a just society where everyone has enough food, clothing, shelter, self-worth, and the leisure time to contemplate the higher spiritual meanings of life. Do we not owe it to each other to face down our fears and act to create this future for our children's children?

Free energy technology is here. It has been here for decades. Communications technology and the Internet have torn the veil of secrecy off of this remarkable fact. People all over the world are starting to build free energy devices for their own use. The bankers and the governments do not want this to happen, but cannot stop it. There will be essentially no major media coverage of what is going on. Tremendous economic instabilities and wars will be used in the near future to distract people from joining the free energy movement.

Western society is in many ways spiraling down toward self-destruction due to the accumulated effects of long-term greed and corruption. The general availability of free energy technology cannot stop this trend. It can only reinforce it. If, however, you have a free energy device, you may be better positioned to support the political/social/economic transition that is underway. The question is, who will ultimately control the emerging world government—the first force or the fourth force?

The last great war is almost upon us. The seeds are planted. After this will come the beginning of a real civilization. Some of us who refuse to fight will survive to see the dawn of the world of free energy. I challenge you to be among the ones who try.

LIST OF RESOURCES:

Books:
Living Energies by Callum Coats
The Free Energy Secrets of Cold Electricity by Peter Lindemann, D.Sc.
Applied Modern 20th Century Aether Science by Dr. Robert Adams
Physics Without Einstein by Dr. Harold Aspden
Secrets of Cold War Technology by Gerry Vassilatos
The Coming Energy Revolution by Jeane Manning

Websites:

http://free-energy.ws/peter-lindemann/ developed by Clear Tech, Inc. and Dr. Peter Lindemann

http://jnaudin.free.fr/ developed by JLN Labs in France

http://www.keelynet.com/ developed by Jerry Decker in the USA

http://www.xogen.ca/ site for super electrolysis technology

http://www.fortunecity.com/greenfield/bp/16/content1.htm excellent site by Geoff Egel, Australia

Resource[i] Article – The World of Free Energy By Peter Lindemann, D.Sc.. Extracted from the Internet @ https://www.wanttoknow.info/freeenergy on January 18, 2019

Section One

Patent and Inventor Suppression

The U.S. Government Has a Secret System for Stalling Patents

Entrepreneurs and established companies alike depend on the U.S. Patent and Trademark Office to legally protect their inventions. But the Patent Office has been using a secret system to withhold the approval of some applications.

Newly released documents reveal that the office, tasked with evaluating and protecting the rights to intellectual property, has a covert system for delaying controversial or inconvenient patents. It's a system that attorneys say, if abused, could function as a way to limit or stomp out emerging companies.

Before today, the program — named the Sensitive Application Warning System (SAWS) — has been mentioned only anecdotally by examiners who work in or with the office, and in a government memo that was leaked in March 2006. However, a new 50-page document obtained by a law firm's Freedom of Information Act request shows the sweeping scope and conflicting interests of this particular set of rules. The law firm behind the request, Kilpatrick Townsend & Stockton LLP, frequently represents major tech companies, including Apple, Google, Twitter, and Oracle.

Delays by design

When an application is submitted for a patent from a major law firm, it usually requires the approval of one or two examiners who work within the Patent Office. For Thomas Franklin, a partner at Kilpatrick Townsend, applications that he prosecutes typically issue as patents 22 months after filing. (Though the USPTO's website estimates the average patent pendency time to be 29.1 months, that figure considers independent applicants who are not represented by powerhouse law firms.)

Any application that is categorized in SAWS, however, is placed in a special type of patent purgatory. SAWS-marked patents must be approved by anywhere from three to nine people and can be delayed for years. There is no official channel to notify an applicant once their patent is placed in the system, and the Patent Office has denied requests to divulge what applications are on the SAWS list.

For instance, three years ago, Franklin submitted an application for a startup that needed to secure crucial patents before it filed for public offering. It was being sued by a large company in the same space and so it paid the Patent Office an extra fee to fast-track the application. Franklin and his client met with the examiner who was in charge of the case and he seemed fairly confident that it would be approved in a timely manner.

But then, out of the blue, the examiner alluded to a special approvals process that would delay the project.

"He at first made me aware of that and said well, 'That's secret, I'm not supposed to say it,' " Franklin told Yahoo Tech. "That's what piqued my interest as a constitutional issue. There's a secret program that they're not supposed to talk about."

To this day, the patent — which Franklin says is for fairly mundane technology — remains in limbo, with little communication as to why or when it will be resolved from the USPTO. Franklin worries that the system could slow innovation in America.

"If it's a startup trying to get traction against an established player, they may not get funding because they don't get the patent in place," he told Yahoo Tech. "It favors the entrenched incumbent in that space over the person with perhaps breakthrough technology."

Patent applications can be placed in the SAWS program for an extensive number of reasons, which Franklin and his associates call "astonishingly vague." This includes applications of "broad" or "pioneering scope," "seemingly frivolous or silly subject matter," or those "dealing with inventions, which, if issued, would potentially generate unwanted media coverage (i.e., news, blogs, forums)."

Whom does SAWS serve?

The final item on this list could include literally any technology but may be particularly aimed at inventions from high-profile companies like Apple, whose patents are often picked apart by dedicated blogs like 9to5Mac or MacRumors.

The obtained documents also list areas of technology that might trigger a patent application to be placed in the SAWS program, including "smartphones," "Internet-enabled systems," and "processes and apparatuses involving Education."

It's these vast categories that Adam Charnes, a partner at the firm, says could raise constitutional issues.

"Presumably, not every smartphone patent application is sent into this program," he told Yahoo Tech. "There's no way they could be. There's too many of them. If the criteria don't give any guidance to the agency in deciding what's in or what's out, it leads to the very arbitrariness that we're concerned about."

Charnes and his colleague Kate Gaudry plan to publish an article for Law360, an online journal, in hopes of informing the public about this program, which they say most companies are wholly unaware of.

UNITED STATES PATENT AND TRADEMARK OFFICE

OFFICE OF THE GENERAL COUNSEL

October 9, 2014

VIA U.S. MAIL
Ms. Kate Gaudry
5503 Charlcote Road
Bethesda, MD 20817

Re: ***Freedom of Information Act (FOIA) Request No. F-15-00004***

Dear Ms. Gaudry:

The United States Patent and Trademark Office (USPTO) FOIA Office received your e-mail dated October 2, 2014, in which you posed the following questions, under the provisions of the Freedom of Information Act, 5 U.S.C. § 552:

> 1) What are the currently identified potential SAWS subject matters (e.g., subject matters of special interest for each technology center and coprs-wide potential SAWS subject matter); and
>
> 2) For applications determined to contain SAWS material, what is the protocol for review of the application? For example, who is involved in searching for prior art, identifying whether a rejection is to be made, determining whether an application is to be allowed, and/or reviewing a draft Act (e.g., Office Action or Notice of Allocance).

Please be aware that the FOIA is a means by which members of the public can obtain Agency records in existence at the time of the request. It cannot provide answers to questions or create new records in order to respond to FOIA requests. See Hudgins v. Internal Revenue Serv., 620 F. Supp. 19, 21 (D.D.C. 1985).

However, the USPTO has identified forty-three (43) pages of records that it believes are responsive to your request. A copy of this material is enclosed.

Your request is considered complete with full disclosure. However, you have the right to appeal this initial determination to the General Counsel, United States Patent and Trademark Office, P.O. Box 1450, Alexandria, VA 22313-1450. An appeal must be received within 30 calendar days from the date of this letter. See 37 C.F.R. § 102.10(a). The appeal must be in writing. You must include a copy of your original request, this letter, and a statement of the reasons why this response is deficient. Both the letter and the envelope must be clearly marked "Freedom of Information Appeal."

"The Patent Office is kind of a funny place," Franklin said. "We have this manual of patenting examining procedure, and it's 1,500 pages long. They tell us everything: all their criterion in excruciating detail. But there is no mention in this huge book of this SAWS program. It's completely off the books."

The USPTO's communication desk did not offer a public comment by time of publication.

Following, are the papers obtained from Kilpatrick Townsend & Stockton LLP's Freedom of Information Act request.

Reference [ii] Article - The U.S. Government Has a Secret System for Stalling Patents. Written By Alyssa Bereznak, National Correspondent, Technology, Yahoo Tech, December 3, 2014 Extracted from the Internet January 2, 2019 at https://finance.yahoo.com/news/the-u-s-government-has-a-secret-system-for-104249688314.html

The Invention Secrecy Act of 1951

Is An Act to provide for the withholding of certain patents that might be detrimental to the national security, and for other purposes. Citation 66 Stat. 3. Date enacted February 1st, 1952. Bill citation Pub. L. 82–256.

The U.S. government has long sought to control the release of new technologies that might threaten the national defense and economic stability of the country. During World War I, Congress authorized the United States Patent and Trademark Office (PTO) to classify certain defense-related patents. This initial effort lasted only for the duration of that war but was reimposed in October 1941 in anticipation of the U.S. entry into World War II. Patent secrecy orders were initially intended to remain effective for two years, beginning on July 1, 1940, but were later extended for the duration of the war.

The Invention Secrecy Act of 1951 made such patent secrecy permanent, though the order to suppress any invention must be renewed each year (except during periods of declared war or national emergency). Under this Act, defense agencies provide the PTO with a classified list of sensitive technologies in the form of the "Patent Security Category Review List" (PSCRL). The decision to classify new inventions under this act is made by "defense agencies" as defined by the President. Generally, these agencies include the Army, Navy, Air Force, National Security Agency (NSA), Department of Energy, and NASA, but even the Justice Department has played this role.

A secrecy order bars the award of a patent, orders that the invention be kept secret, restricts the filing of foreign patents, and specifies procedures to prevent disclosure of ideas contained in the application. The only way an inventor can avoid the risk of such imposed secrecy is to forgo patent protection.

By the end of fiscal year 1991, the number of patent secrecy orders stood at 6,193. Many such orders were imposed on individuals and organizations working without government support. This number shrank for each fiscal year thereafter, until 2002. Since 2002, the number of secrecy orders has grown, with 5,002 secrecy orders in effect at the end of fiscal year 2007.

The types of inventions classified under this Act are themselves secret, but most of the inventions which are now no longer secret but once were secret have been in areas with high military significance, such as cryptography and weapons development. Many inventions relating to power generating have also had their patents denied under this act.

Action to Declassify Secret Energy Patents

A Proposed Bill to Support New Energy Sources

Las Vegas-area Congresswoman Shelley Berkley announced last August, at a meeting in Las Vegas, a bill for the Small Business Administration to annually provide 750 million dollars in energy loan guarantees (see Berkley Reveals Green Energy Solutions).

In response to Berkley's subsequent request for responses from the audience, I explained that the U.S. Patent Office has a nine-member committee which screens patents for national security implications, and that a hidden purpose of this committee is to lock up energy-related patents which could threaten the fossil-fuel monopolies and the power grid (nuclear, coal, etc.). When an inventor has his or her energy patent classified, I explained, the inventor faces 20 years in prison for working on or publicizing the invention.

I told Berkley and the audience that a Space Energy Journal article by Don Kelly claimed some years ago that there are now 3000 energy patents which have been classified, and that I considered it highly likely that there were some potentially very productive new sources of energy locked up in those 3000 patents.

I asked Berkley to add a section to her bill to establish some kind of mechanism for going back and reviewing all these classified energy patents for possible release and government support for their development.

Berkley seemed favorably responsive. I offered to help her write the section, and was referred to a staff member.

Below are a few stories to give readers who are unfamiliar with energy invention suppression a feeling for the ruthlessness and unfairness of this dirty, greedy business.[1]

Energy Patent Suppression: Seven Cautionary Tales

Story No. 1

Neil Schmidt stopped in my office on May 12, 1995, to tell me this story about his invention of a hydraulic wind turbine. At this time, he had been living in Las Vegas for seven years. But nine years before, he was living in the Seattle area, and had visited a federal Small Business Administration office there to apply for financial aid.

The following day, he received a telephone call. He was told his device would not work, and not to bother with it.

He had provided the SBA office with hardly more than a sketch, so a hot argument erupted which lasted half an hour. The caller ended up hinting death to Neil if he didn't stop working on his invention.

A couple of days later, Neil went back to the office and walked around unsuccessfully trying to identify the voice he had heard on the telephone.

Neil also has heard that another energy inventor living near his Washington residence had been shot in the head and blinded.

Story No. 2

When one of my inventor friends was a young man, he invented a 90-mpg carburetor. He was paid a visit by four men, including one from Standard Oil and two wearing U.S. Marshal's uniforms.

They told him that if he ever made another carburetor, they would kill him, his wife, and two young children. He was quickly persuaded that his life wasn't worth a "damn carburetor."

He happened to think to memorize the badge numbers of the two "US Marshals," and so had an attorney in Washington, DC, check with the US Marshal's office. They had no record of the two badge numbers.

Story No. 3

In the 1930s, Henry T. Moray was refused a patent on his cold semiconductor cathode, allegedly because the patent examiner couldn't understand how it could emit electrons. Twenty years later, the development of the transistor proved that his device could actually have worked.

Then in 1940, Moray demonstrated before members of the Public Utilities Commission (PUC) his free energy generator, which gave a continuous output of 250,000 volts with no apparent input. The next day he was found shot in his lab, and all of his notes and device were stolen.

Currently, the Cosray Research Institute of Salt Lake City has been collecting Moray's papers from old acquaintances and attempting to duplicate the device.

Story No. 4

In May 1974, John Andrews, a Portuguese chemist, demonstrated a water-to-gas additive before Navy officials. The additive allowed ordinary water to be added to gasoline without decreasing the combustibility of the gas. It would have driven the cost of gasoline down to two cents per gallon.

When U.S. Navy officials finally went to his lab to negotiate for the formula, they found Andrews missing and his lab ransacked.

Story No. 5

In July 1974, the Los Angeles District Attorney, acting under orders from the "top," raided the assembly plant of Ed Gray and confiscated his prototype generator, as well as his plans and records. They brought numerous false charges against him, and defied all attempts by Gray's lawyers to get the confiscated materials returned.

Gray's opponents have driven him into bankruptcy (Generator, Free Energy, Patent No. 38,905,480).

Story No. 6

On September 3, 1978, a U.S. Government agency restricted development of a CB radio and telephone privacy device. The device had been invented by four men — Carl Kicolai, William Raike, Carl Quale, and David Miller, of Seattle, Washington. It prevents eavesdropping on CB radio and telephone transmissions for nonmilitary, nongovernment use.

The patent was applied for and refused. The National Security Agency was quick to issue a secrecy order against them,[2] preventing them from even talking about the invention without risking a jail term.

The order was received six months after a patent was filed for the device. The Patent Office wrote: "Your application . . . has been found to contain subject matter, the unauthorized disclosure of which might be detrimental to national security."

Story No. 7

Most people believe that it is impossible to build a motor run solely by magnets. Howard Johnson spent six years fighting the Patent Office to accept the reality of his magnet motor. They finally granted him U.S. Patent 4,151,431. There are dozens of other magnet motor patents.

Many achievements in developing free energy from magnetism have occurred in obscure basement workshops all over the country. In most cases, these achievements remain the secret of the inventor, who fears reprisal.

Engineers at Hitachi Magnetics Corporation have publicly admitted that a motor run by magnets is feasible and logical. But the politics of the matter make it impossible for them to pursue developing a magnet motor, or any device that would compete with the energy cartels.

Recent Developments: Reviving the OTA

OTA, or Office of Technology Assessment, is an arm of Congress that formerly — from 1972 to 1995 — provided comprehensive assessments of a wide range of scientific and technical issues.

In August 2001, I learned that Rush Holt (D-NJ) has introduced legislation in the House of Representatives proposing to return the OTA to active status (OTA — see cnie.org/Updates/106.htm).

During its years of operation, OTA supplied Congress with background papers, briefings, and testimony, but was best known for its in-depth, objective reports of complex scientific and technical issues. These reports took one to two years to produce and could be requested by the chairman of any Congressional committee.

OTA's Environment Program addressed areas including environmental health and risk management, the use of renewable resources, and pollution prevention, control and remediation. Its reports covered topics such as agriculture, biological diversity, air and water pollution, the effects of weather and climate change, management of solid, hazardous, and nuclear wastes, risk assessment methods and policy, and public land management.

Resurrecting OTA was suggested in Recommendations for Improving the Scientific Basis for Environmental Decision Making, a report from the first national conference of the National Council for Science and the Environment (NCSE). NCSE suggested reviving the Office of Technology Assessment as a change that could "significantly improve efficiency and communication among scientists and between scientists and decision makers."

The full text of this report can be found online at NCSEonline.org.

National Public Radio's July 18 broadcast of "All Things Considered" included a segment on reopening OTA. During this piece, Rep. Sherwood Boehlert (R-NY), chairman of the House Science Committee, and Rep. Vernon Ehlers (R-MI), a physicist by training, lauded the work of OTA and their "wonderful, detailed, long-term studies."

However, Newt Gingrich, Speaker of the House of Representatives when OTA was eliminated, noted that many conservatives felt that OTA "was used by liberals to cover up political ideology with a gloss of science."

Rep. Holt's bill, H.R. 2148, would revive OTA by simply reenacting its 1972 formative legislation and authorizing a budget of 20 million dollars per year for five years. This is the same level of funding OTA received in 1994. Currently, this legislation has 50 cosponsors.[3].

To express your opinion on H.R. 2148 and the revival of OTA, contact your Representative and Senators at:

Honorable [insert name], United States House of Representatives, Washington, DC 20515

Senator [insert name], United States Senate, Washington, DC 20510

For details about OTA, including an archive of reports, see www.princeton.edu/ota, or contact Kevin Hutton, Webmaster, National Council for Science and the Environment, 1725 K St. NW, Suite 212, Washington, DC 20006, website cnie.org.

It's Probably Closer to 4000

Thomas Valone, of the Integrity Research Institute (iri@erols.com) wrote to me saying that my suggestion to Congresswoman Berkley was "an answer to a prayer."

Integrity Research Institute (IntegrityResearch.org) and the Disclosure Project[4] both support your suggestion.

I will be happy to cooperate with your requests (see "How You Can Help," below), and even meet with staff person Guiton at his convenience.

As a former Patent Examiner, I can tell you that the number of "securitized" patents in the vault at the Patent Office (Park 5 Bldg.) is closer to 4000 or more. They [applicants] never receive a patent number, and the inventor is rarely, if ever, compensated by the government for use of the invention.

Since you have taken the initiative to start this process, we can help by contacting our Congresspeople and asking them to send letters to Congresswoman Berkley in support.

I also suggest that we all should ask our Congresspersons to contact the three committees that are pertinent to this subject: Environment and Public Works (headed by James Jeffords); Energy and Natural Resources (headed by Jeff Bingaman); and Commerce (headed by Ernest Hollings).

Help from Down Under

Concerning the proposal of an energy bill to declassify patents, Australia's Trevor Osborne (wharmony@iinet.net.au) wrote me that he has sent out an email to about 100 opinion leaders who might help perpetuate the momentum.

"So the momentum is gaining speed," Osborne wrote. "First, we had the Disclosure Project. Now we have the proposed Energy Bill to Declassify Energy Patents. So please do your best to pass this news on to anyone who may be interested in writing to and/or phoning their Congressperson to offer their support to this bill."

Meanwhile on Capitol Hill

The energy bill that was actually passed by the House was not about strengthening America's strategic energy position. Rather, it was about something much simpler — a gigantic giveaway to the energy industry.

The dishonesty began with the bill's very first sentence: "To enhance energy conservation, research, and development. . ." In fact, the bill not only ignored conservation, it actually subverted it, epitomizing Vice President Cheney's view that "conservation is for wimps."

The bill, among other things, extends a loophole for automobiles that can also run on ethanol but rarely do; this means that overall vehicle fuel efficiency would probably be reduced. Yet 36 Democrats joined Republicans in voting for this legislation, giving President Bush a comfortable margin of victory — and, as with the tax cut, bipartisan cover.

Not surprisingly, it was mostly Democrats from auto-producing and energy states who supported the bill. These included John D. Dingell (Michigan), Brad Carson (Oklahoma) and Chet Edwards (Texas), as well as Californians Joe Baca (Rialto) and Calvin M. Dooley (Visalia).

The bill is a natural target for Democrats. For one thing, the links between campaign contributions and the bill's provisions could hardly be clearer. In an analysis released by Rep. Henry A. Waxman (D-Los Angeles) called "Hitting the Jackpot," the rewards for industry are outlined.

The report mischievously observes that the cumulative value of the 2000 election-campaign contributions by the coal, oil and gas, nuclear, and utility industries — the lion's share of which went to the GOP — was 69.5 million dollars. But the total value of the tax breaks and subsidies is 36.4 billion dollars. The rate of return on these campaign contributions was thus a whopping 52,200 percent!

Specifically, the coal mining industry, which contributed 3.8 million dollars, would receive 1.1 billion dollars in direct subsidies in the next three years, and an extra 1.4 billion dollars over the next seven years.

Typical of the bill is that government would assume the industry's costs for applying to mine coal on federal lands.

The oil industry, which contributed 33.3 million dollars in 2000, would get tax breaks worth 12.8 billion dollars, according to the Joint Committee on Taxation. For instance, it would no longer have to pay royalties for oil and gas lease sales.

The only good news is that the Senate is not likely to pass the energy bill unamended.

How You Can Help

I can see several ways you can help me declassify energy patents by focusing a public spotlight on the Patent Office's patent secrecy performance.

First, you can help me by providing references to back up my claims concerning the 9-member screening committee, the 20-year jail sentence for breaking a "gag order," and the number and nature of the classified patents. I will in turn send these to Mark Guiton, legislative director in the U.S. House of Representatives.[5]

Second, it will also help if you can provide me with name and contact information for an expert on classified patents. I can then suggest to Mark Guiton that he go to this expert for information and advice.

Third, if you can, please give thought to the mechanism that might be incorporated into the new section of the legislation, specifying exactly how the patents are to be reviewed and declassified. One crucial feature would involve how people are selected for the review agency to assure that they will be honest, rather than "toadies" for fossil fuel and nuclear interests.

I feel that it would be of value to hold a public seminar in the Washington, DC, area on the subject — perhaps in a Patent Office auditorium. Any way in which you can lend support to such a seminar would be welcome.

And it will assist my own endeavors if someone could email me a copy of the actual form the Patent Office uses when directing an inventor to stop working on an invention (vman@skylink.net).

In Conclusion

I think Congresswoman Berkley deserves high praise for introducing an unusually progressive energy bill. I urge you all to write your congresspeople and senators to support a bill which would include a provision for systematically reviewing all classified energy patents for release.

Footnotes:

Please note that I have not independently verified the truthfulness of these stories.

Called "gag orders," defiance risks a 20-year prison term.

Cosponsors of the OTA-revival bill include Thomas H. Allen (D-ME), Tammy Baldwin (D-WI), Judy Biggert (R-IL), Earl Blumenauer (D-OR), Sherwood Boehlert (R-NY), Rick Boucher (D-VA), Lois Capps (D-CA), Michael Capuano (D-MA), Wm. Lacy Clay (D-MO), William J. Coyne (D-PA), Joseph Crowley (D-NY), John D. Dingell (D-MI), Lloyd Doggett (D-TX), Michael F. Doyle (D-PA), Vernon J. Ehlers (R-MI), Eliot L. Engel (D-NY), Bob Filner (D-CA), Wayne Gilchrest (R-MD), James Greenwood (R-PA), Tony P. Hall (D-OH), Jane Harman (D-CA), Joseph M. Hoeffel (D-PA), Rush Holt (D-NJ), Michael M. Honda (D-CA), Amo Houghton (R-NY), Steny H. Hoyer (D-MD), Dale E. Kildee (D-MI), Ron Kind (D-WI), John J. LaFalce (D-NY), John Lewis (D-GA), Zoe Lofgren (D-CA), Edward J. Markey (D-MA), Karen McCarthy (D-MO), Jim McDermott (D-WA), Cynthia McKinney (D-GA), James P. Moran (D-VA), Constance Morella (R-MD), Jerrold Nadler (D-NY), Richard E. Neal (D-MA), John W. Olver (D-MA), Bill Pascrell, Jr. (D-NJ), Donald M. Payne (D-NJ), David E. Price (D-NC), Tom Sawyer (D-OH), Christopher Shays (R-CT), Louise McIntosh Slaughter (D-NY), Adam Smith (D-WA), Vic Snyder (D-AR), Hilda L. Solis (D-CA), and David Wu (D-OR).

See disclosureproject.org. See also this month's Spirit of Ma'at article titled Secrets From the Stars.

His email is Mark.Guiton@Mail.House.Gov. He also can be reached: by mail at U.S. House of Representatives, 439 Cannon Building, Washington, DC 20515; by phone at 202-225-5965; and by fax at 202-225-3119.

Reference[iii] Article - Action to Declassify Secret Energy Patents written by Gary Vesperman. Extracted from the Internet January 18, 2019 @ http://www.spiritofmaat.com/archive/feb2/prns/vesprman.htm. Gary Vesperman is a tireless writer and activist with knowledge in a vast area of subjects of interest and concern to modern technology and our way of life. His mailing address is 3123 Trueno Road, Henderson, NV 89014-3142, phone 702-435-7947. He can be reached via email at vman@skylink.net.

Section Two

Media Manipulations

Leading Journalists Expose Major News Cover-ups by Media Corporations

This is a book review. In this engaging book, 20 top journalists detail their personal experiences of trying to report major stories only to have them systematically shut down by their own media corporations. The stories include the CIA's involvement in introducing crack cocaine into LA, the accidental shooting down of TWA flight 800 by the military, outrageous election fraud Nov. 2000 in Florida, and lots more. You will learn how American mass media is strictly controlled by an amazingly small cast of power brokers. Though we do have some freedom of the press, you will learn that the biggest, most important issues of our time are rarely covered by our media. After reading this book, you will understand how some of the huge cover-ups mentioned on this resource list are managed. Chapters 3, 13, 14, and 18 are particularly fascinating.

The riveting excerpts below from the revealing accounts of 20 award-winning journalists in the highly acclaimed book Into the Buzzsaw are essential reading for all who support democracy. These courageous writers were prevented by corporate ownership of the mass media from reporting major news stories. Some were even fired. They have won numerous awards, including several Emmys and a Pulitzer. If you want to understand how the media powerfully influences our understanding of the world, read this book! https://www.amazon.com/exec/obidos/tg/detail/-/1591022304/ref=ord_cart_shr/102-2931196-4693767?tag=wanttinfo-20&_encoding=UTF8&m=ATVPDKIKX0DER&v=glance

Jane Akre spent 20 years as a network and local TV reporter for news and mass media operations throughout the country. She and her husband, investigative reporter Steve Wilson, were awarded the Goldman Environmental Prize for their struggle with mass media ownership related here.

By February 1997 our story was ready to air. It attempted to answer some troubling questions: Why had Monsanto sued two small dairies to prevent them from labeling their milk as coming from cows not injected with [growth hormone rBGH]? Why had two Canadian health regulators claimed that their jobs were threatened – and then said Monsanto offered them a bribe to give fast-track approval to the drug? Why did Florida supermarkets break their much-publicized promise that milk in the dairy case would not come from hormone-treated cows? And why was the US the only major industrialized nation to approve this controversial genetically engineered hormone? (p. 211)

Station managers were so proud of our work that they saturated virtually every Tampa Bay radio station with thousands of dollars' worth of ads urging viewers to watch what we'd uncovered about "The Mystery in Your Milk." But then, our Fox managers' pride turned to panic. [Monsanto lawyer] John Walsh wrote that some points of the story "clearly contain the elements of defamatory statements which, if repeated in a broadcast, could lead to serious damage to Monsanto and dire consequences for Fox News." (pp. 211-213)

It was not long after our [unsuccessful] struggle to air an honest report had begun that Fox fired both the news director and the general manager. The new general manager, Dave Boylan, explained that if we didn't agree to changes that Monsanto and Fox lawyers were insisting upon, we'd be fired for insubordination within 48 hours. We pleaded with Dave to look at the facts we'd uncovered, many of which conclusively disproved Monsanto's claims. We reminded him of the importance of the facts about a basic food most of our viewers consume and feed to their children daily. His reply: "We paid $3 billion dollars for these TV stations. We'll tell you what the news is. The news is what we say it is!" Steve [the author's husband and coworker] was firm but respectful when he made it clear we would neither lie nor distort any part of the story. (pp. 213-215)

[The Dairy Coalition's director] took great pride in bragging that the Coalition "snowed the station with paperwork and pressure to have the story killed." Fox threatened our job every time we resisted the dozens of changes that would sanitize the story and fill it with lies and distortions. [Fox lawyer] Forest finally leveled with us. "You guys don't get it. It doesn't matter whether the facts are true. This story isn't worth a couple of hundred thousand dollars to go up against Monsanto." (pp. 217, 218)

Fox's general manager presented us with an agreement that would give us a full year of salaries and benefits worth $200,000 in no-show "consulting jobs," but with strings attached: no mention of how Fox covered up the story and no opportunity to ever expose the facts Fox refused to air. We turned down this second hush money offer. We were both finally fired, allegedly for "no cause." (p. 219)

The controversy over rBGH has traveled recently to Canada and the European Union, both of which decided to reject the drug for use in those countries. (p. 236)

For a revealing 10-minute video clip of this astounding case, click here. For updates on their lawsuit, see the Ms. Akre and Mr. Wilson's website at http://www.foxbghsuit.com.

Dan Rather [was] the anchor and managing editor of CBS Evening News and correspondent for 60 Minutes II. In his more than 30 years at CBS, he received almost every honor in broadcast journalism, including several Emmy Awards, a Peabody Award, and citations from scholarly, professional, and charitable organizations. This is an excerpt from an interview originally aired on BBC Newsnight on May 16, 2002.

Access was extremely limited to the press during the time of September 11th, and ever since then [has been] limited in a way that is unprecedented in American journalism. There was a full understanding of why access was so limited during that time. [However] in the weeks and months that followed September 11th, the federal government began to take an unprecedented attitude about the access of American journalists to the war. What's particularly troubling is that what's being done is in direct variance with the Pentagon's stated policy [of] maximum access and maximum information consistent with national security. What's going on is a belief that you can manipulate communicable trust between the leadership and the led. The way you do that is you don't let the press in anywhere (p. 36-38).

Access to the [Iraq] war is extremely limited. The fiercer the combat, the more the access is limited, [including] access to information. I would say that overwhelmingly the limiting of access to information has much more to do with the determination to be seen as conducting the war errorlessly than it does with any sense of national security (p. 40).

None of us in journalism have asked questions strongly enough about limiting access and information for reasons other than national security. It's unpatriotic not to ask questions. Anybody in American journalism who tells you that he or she has not felt this pressure [not to ask tough questions] is either kidding themselves or trying to deceive you (p. 39-40)

What we're talking about here is a form of self-censorship. Self-censorship is a real and present danger to journalists at every level and on a lot of different kinds of stories. Before the war, before September 11th, fear ruled every newsroom in the country in some important ways – fear if we don't dumb it down, if we don't tart it up, if we don't go to the trivial at the expense of the important, we're not going to be publishing a newspaper or magazine. We're not going to be on the air. The ratings will eat us up. (p. 41-42).

There was a time in South Africa when people would put flaming tires around people's necks if they dissented. In some ways the fear [now in the U.S.] is that you'll have a flaming tire of lack of patriotism put around your neck. It's that fear that keeps journalists from asking the tough questions. And I am humbled to say, I do not except myself from this criticism (p. 42).

Monika Jensen-Stevenson is a former Emmy-winning producer for 60 Minutes. The Vietnam Veterans Coalition awarded her the Vietnam Veterans National Medal.

Marine Private Robert R. Garwood – fourteen years a prisoner of the communist Vietnamese – was found guilty of collaboration with the enemy in the longest court-martial in United States history. I first heard of Garwood in 1979. Wire reports referred to him as a defector whom the US government was charging with being a traitor. At the end of the court-martial, there seemed no question that Garwood was a monstrous traitor. (pp. 255, 256)

In 1985, Garwood was speaking publicly about something that had never made the news during his court-martial. The Wall Street Journal reported he said that he knew firsthand of other American prisoners in Vietnam long after the war was over. He was supported by Vietnam combat veterans whose war records were impeccable. These veterans told a story vastly different from what was made public during the court-martial and one that was intimately tied to another 60 Minutes story I was working on – "Dead or Alive?" The title referred to Vietnam POW/MIAs [Prisoners Of War/Missing In Action]. (p. 256)

My sources included outstanding experts like former head of the Defense Intelligence Agency General Eugene Tighe and returned POWs like Captain Red McDaniel, who held the Navy's top award for bravery, had commanded the aircraft carrier Lexington, and was director of liaison on Capitol Hill for the Navy and Marine Corps. With such advocates providing back up, it was hard not to consider the possibility that prisoners (some 3,500) had in fact been kept by the Vietnamese communists as hostages to make sure the US would pay the more than $3 billion in war reparations that Nixon had promised before his fall from grace. Particularly compelling was the fact that of the 300 prisoners known to be held in Laos, not one was released for homecoming in 1973. (p. 256)

Initially held back to ensure the US would fulfill its secret promise to pay reparation monies, by 1979 American POWs had become worthless pawns. The US had not paid the promised monies and had no intention of paying in the future. (p. 263)

Ms. Jensen-Stevenson's book on this topic, Kiss the Boys Goodbye, is available at amazon.com.

Kristina Borjesson has been an independent producer and writer for almost 20 years. Among her many accomplishments besides editing this volume, she worked at CBS network where she won an Emmy and a Murrow Award for her investigative reporting on "CBS Reports: Legacy of Shame" with Dan Rather and Randall Pinkston.

You don't choose to have the kind of experience I had while trying to report on the demise of TWA Flight 800. You fall into it. At CBS, I'd recently picked up an Emmy for investigative reporting when I was assigned to investigate the crash. I had no idea that my life would be turned upside down and inside out – that I'd be assigned to walk into what I now call "the buzzsaw." (p. 284)

The buzzsaw is what can rip through you when you try to investigate or expose anything this country's large institutions – be they corporate or government – want kept under wraps. The system fights back with official lies, disinformation, and stonewalling. Your phone starts acting funny. Strange people call you at strange hours to give you strange information. The FBI calls you. Your car is broken into and the thief takes your computer and your reporter's notebook and leaves everything else behind. You feel like you're being followed everywhere you go. (p. 284)

Pierre Salinger announced to the world on November 8, 1996, that he'd received documents from French intelligence proving that a US Navy missile had accidentally downed [TWA Flight 800]. That same day, FBI's Jim Kallstrom called a press conference to deny Salinger's allegations. [At the press conference,] Kallstrom rattled off a prepared speech, and then it was time for questions. A man raised his hand and asked why the Navy was involved in the recovery and investigation while a possible suspect. Kallstrom's response was immediate; "Remove him!" he yelled. Two men leapt over to the questioner and grabbed him by the arms. There was a momentary chill in the air after the guy had been dragged out of the room. Kallstrom acted as if nothing had happened. (pp. 290, 291)

A few weeks after the FBI's visit to CBS, I received my walking papers. Law enforcement consultant Paul Ragonese eventually got his walking papers, too. Ragonese was replaced by none other than the FBI's TWA 800 task force chief, James Kallstrom. (p. 307)

Ms. Borjesson compiled Into the Buzzsaw, the book from which this summary is made.

Greg Palast writes for the Guardian and Observer newspapers of London and reports for the BBC's Newsnight. Palast abandoned hopes of working in America when mainstream press failed to report on his groundbreaking exposes known for stripping bare abuses.

In the months leading up to the November [2000] balloting, Florida Governor Jeb Bush and his secretary of state, Katherine Harris, ordered local elections supervisors to purge 58,000 voters from registries on the grounds they were felons not entitled to vote in Florida. As it turns out, only a handful of these voters were felons. The voters were [about 54%] African Americans, and most of the others were white and Hispanic Democrats. [Several] weeks after the election, this extraordinary news ran on page one of the country's leading paper. Unfortunately, it was in the wrong country: Britain. In the USA, it was not covered. It was given big network TV coverage. But again, it was on the wrong continent – on BBC TV, London. (pp. 195, 196)

The office of the governor [also] illegally ordered the removal of felons from the voter rolls – real felons – but with the right to vote under Florida law. As a result, 50,000 of these voters could not vote. The fact that 90% of these voters were Democrats should have made it news because this maneuver alone more than accounted for Bush's victory. (pp. 197-200)

In February 2001, I took my BBC film crew to Florida, having unearthed a page marked "secret" and "confidential" from the company the state had hired to make up the list of names to purge from voter rolls. I took my camera crew into an agreed interview with Jeb Bush's director of the Department of Elections. When I pulled out the confidential sheet, Bush's man ripped off the microphone and did the fifty-yard dash, locking himself in his office, all in front of our cameras. It was killer television and wowed the British viewers. We even ran a confession from the company. Newsworthy for the USA? Apparently not. (pp. 202, 203)

A group of well-placed sources told my BBC team that before Sept. 11th the US government had turned away evidence of Saudi billionaires funding bin Laden's network. We got our hands on documents that backed up the story that FBI and CIA investigations had been slowed by the Clinton administration, then killed by Bush Jr.'s. The story made top of the news – in Britain. In the US, one TV reporter picked up the report. He was called, he says, by network chiefs, and told to go no further. He didn't. (p. 205)

For Mr. Palast's website, see http://www.gregpalast.com.

Michael Levine is a 25-year veteran of the DEA turned best-selling author and journalist. His articles and interviews on the drug war have been published in numerous national newspapers and magazines, including the New York Times, Los Angeles Times, USA Today, and Esquire.

When Nixon first declared war on drugs in 1971, there were fewer than 500,000 hard-core addicts in the nation, most of whom were addicted to heroin. Three decades later, despite the expenditure of $1 trillion in tax dollars, the number of hard-core addicts is shortly expected to exceed five million. Our nation has become the supermarket of the drug world, with a wider variety and bigger supply of drugs at cheaper prices than ever before. The problem now not only affects every town on the map, but it is difficult to find a family anywhere that is not somehow affected. (pp. 158, 159)

The Chang Mai factory the CIA prevented me from destroying was the source of massive amounts of heroin being smuggled into the US in the bodies and body bags of GIs killed in Vietnam. (p. 165)

My unit, the Hard Narcotics Smuggling Squad, was charged with investigating all heroin and cocaine smuggling through the Port of New York. My unit became involved in investigating every major smuggling operation known to law enforcement. We could not avoid witnessing the CIA protecting major drug dealers. Not a single important source in Southeast Asia was ever indicted by US law enforcement. This was no accident. Case after case was killed by CIA and State Department intervention and there wasn't a damned thing we could do about it. CIA-owned airlines like Air America were being used to ferry drugs throughout Southeast Asia, allegedly to support our "allies." CIA banking operations were used to launder drug money. (pp. 165, 166)

In 1972, I was assigned to assist in a major international drug case involving top Panamanian government officials who were using diplomatic passports to smuggle large quantities of heroin and other drugs into the US. The name Manuel Noriega surfaced prominently in the investigation. Surfacing right behind Noriega was the CIA to protect him from US law enforcement. As head of the CIA, Bush authorized a salary for Manuel Noriega as a CIA asset, while the dictator was listed in as many as 40 DEA computer files as a drug dealer. (pp. 166, 167)

The CIA and the Department of State were protecting more and more politically powerful drug traffickers around the world: the Mujihadeen in Afghanistan, the Bolivian cocaine cartels, the top levels of Mexican government, Nicaraguan Contras, Colombian drug dealers and politicians, and others. Media's duties, as I experienced firsthand, were twofold: first, to keep quiet about the gush of drugs that was allowed to flow unimpeded into the US; second, to divert the public's attention by shilling them into believing the drug war was legitimate by falsely presenting the few trickles we were permitted to indict as though they were major "victories," when in fact we were doing nothing more than getting rid of the inefficient competitors of CIA assets. (pp. 166, 167)

On July 17, 1980, drug traffickers actually took control of a nation. Bolivia at the time [was] the source of virtually 100% of the cocaine entering the US. CIA-recruited mercenaries and drug traffickers unseated Bolivia's democratically elected president, a leftist whom the US government didn't want in power. Immediately after the coup, cocaine production increased massively, until it soon outstripped supply. This was the true beginning of the crack "plague." (pp. 167, 168)

The CIA along with the State and Justice Departments had to combine forces to protect their drug-dealing assets by destroying a DEA investigation. How do I know? I was the inside source. I sat down at my desk in the American embassy and wrote the kind of letter that I never myself imagined ever writing. I detailed three pages typewritten on official US embassy stationary—enough evidence of my charges to feed a wolf pack of investigative journalists. I also expressed my willingness to be a quotable source. I addressed it directly to Strasser and Rohter, care of Newsweek. Two sleepless weeks later, I was still sitting in my embassy office staring at the phone. Three weeks later, it rang. It was DEA's internal security. They were calling me to notify me that I was under investigation. I had been falsely accused of everything from black-marketing to having sex with a married female DEA agent. The investigation would wreak havoc with my life for the next four years. (pp. 168-171)

In one glaring case, an associate of mine was sent into Honduras to open a DEA office in Tegucigalpa. Within months he had documented as much as 50 tons of cocaine being smuggled into the US by Honduran military people who were supporting the Contras. This was enough cocaine to fill a third of US demand. What was the DEA response? They closed the office. (p. 175)

Sometime in 1990, US Customs intercepted a ton of cocaine being smuggled through Miami International Airport. A Customs and DEA investigation quickly revealed that the smugglers were the Venezuelan National Guard headed by General Guillen, a CIA "asset" who claimed that he had been operating under CIA orders and protection. The CIA soon admitted that this was true. If the CIA is good at anything, it is the complete control of

American mass media. So secure are they in their ability to manipulate the mass media that they even brag about it in their own in-house memos. The New York Times had the story almost immediately in 1990 and did not print it until 1993. It finally became news that was "fit to print" when the Times learned that 60 Minutes also had the story and was actually going to run it. The highlight of the 60 Minutes piece is when the administrator of the DEA, Federal Judge Robert Bonner, tells Mike Wallace, "There is no other way to put it, Mike, [what the CIA did] is drug smuggling. It's illegal [author's emphasis]." (pp. 188, 189)

The fact is - and you can read it yourself in the federal court records - that seven months before the attempt to blow up the World Trade Center in 1993, the FBI had a paid informant, Emad Salem, who had infiltrated the bombers and had told the FBI of their plans to blow up the twin towers. Without notifying the NYPD or anyone else, an FBI supervisor "fired" Salem, who was making $500 a week for his work. After the bomb went off, the FBI hired Salem back and paid him $1.5 million to help them track down the bombers. But that's not all the FBI missed. When they finally did catch the actual bomber, Ramzi Yousef (a man trained with CIA funds during the Russia-Afghanistan war), the FBI found information on his personal computer about plans to use hijacked American jetliners as fuel-laden missiles. The FBI ignored this information, too. (p. 191)

Learn about Mr. Levine's books and radio show at http://www.expertwitnessradio.org.

Gary Webb was an investigative reporter for 19 years. He was one of six reporters to win a 1990 Pulitzer Prize for reporting on northern California's 1989 earthquake. He also received the 1997 Media Hero award, and in 1996 was named Journalist of the Year by the Bay Area Society of Professional Journalists. He worked on several newspapers until being forced out of his job after the San Jose Mercury News retracted their support for the Dark Alliance story discussed below.

In 1996, I wrote a series of stories, entitled Dark Alliance, that began this way: For the better part of a decade, a Bay Area drug ring sold tons of cocaine to the Crips and Bloods street gangs of LA and funneled millions in drug profits to a Latin American guerilla army run by the CIA. The cocaine that flooded in helped spark a crack explosion in urban America. It is one of the most bizarre alliances in modern history - the union of a US-backed army attempting to overthrow a socialist government and the Uzi-toting "gangstas" of Los Angeles. (p. 143)

In December 1995, I wrote a lengthy memo to my editors, advising them of what my Nicaraguan colleague and I had found: With the help of recently declassified documents, FBI reports, DEA undercover tapes, as well as interviews with some of the key participants, we will show how a CIA-linked drug and stolen car network provided weapons and tons of high-grade, dirt cheap cocaine to the very person who spread crack through LA and from there into the hinterlands. A bizarre bond between an elusive CIA operative and a brilliant car thief from LA's ghettos touched off a social phenomenon – crack and gang-power – that changed our lives. The day these two men met was literally ground zero for California's crack explosion. This is also the story of how an ill-planned foreign policy adventure – the CIA's "secret" war in Nicaragua – boomeranged back to the streets of America, in the long run doing more damage to us than to our "enemies" in Central America. We have compelling evidence that the kingpins of this cocaine ring enjoyed a unique relationship with the US government that has continued to this day. (pp. 145-146)

The story was developing a political momentum all of its own, and it was happening despite a virtual news blackout from the mass media. Ultimately, it was public pressure that forced the national newspapers into the fray. In Washington, black media outlets were ridiculing the Post for its silence. [In] October and November, the Washington Post, New York Times, and Los Angeles Times published lengthy stories about the CIA drug issue, but spent precious little time exploring the CIA's activities. Instead, my reporting and I became the focus of their scrutiny. The official conclusion reached by all three papers: Much ado about nothing. No story here. The series was "flawed." It was remarkable [Mercury News editor] Ceppos, wrote, that the four Post reporters assigned to debunk the series "could not find a single significant factual error." (pp. 149-152)

At my editor's request, I wrote another series following up on the first three parts: a package of four stories to run over two days. They never began to edit them. Instead, I found myself involved in hours-long conversations with editors that bordered on surreal. A few months later, the Mercury News officially backed away from Dark Alliance, publishing a long column by Jerry Ceppos apologizing for "shortcomings" in the series. The New York Times hailed Ceppos for setting a brave new standard for dealing with "egregious errors" and splashed his apology on their front page, the first time the series had ever been mentioned there. I quit the Mercury News after that. (p. 153)

The CIA's knowledge and involvement had been far greater than I'd ever imagined. Agents and officials of the DEA had protected the traffickers from arrest, something I'd not been allowed to print. At the start of the Contra war, the CIA and Justice Department had worked out an unusual agreement that permitted the CIA not to have to report allegations of drug trafficking by its agents to the Justice Department. It was a curious loophole in the law, to say the least. (p. 154)

The Mercury News had broken the rules and used the Internet to get in by the back door, leaving the big papers momentarily embarrassed. It forced them to readdress an issue they'd much rather have forgotten. By turning on the Mercury News, the big boys were reminding the rest of the flock who really runs the newspaper business, Internet or no Internet, and the extent to which they will go to protect that power, even if it meant rearranging reality to suit them. (p. 155)

Do we have a free press today? Sure we do. It's free to report all the sex scandals it wants, all the stock market news we can handle, every new health fad that comes down the pike, and every celebrity marriage or divorce that happens. But when it comes to the real down and dirty stuff – stories like Tailwind, the October Surprise, the El Mozote massacre, corporate corruption, or CIA involvement in drug trafficking – that's where we begin to see the limits of our freedoms. In today's mass media environment, sadly, such stories are not even open for discussion. Back in 1938, when fascism was sweeping Europe, legendary investigative reporter George Seldes observed that "it is possible to fool all the people all the time – when government and press cooperate." Unfortunately, we have reached that point. (p. 156)

See Mr. Webb's riveting book Dark Alliance on amazon.com. He was found dead in December 2004. It was strangely declared a suicide even though public reports stated that he had not one, but two bullets in his head.

John Kelly is first author with Phillip Wearne of Tainting Evidence: Inside the Scandals at the FBI Crime Lab, which was nominated for a Pulitzer Prize. It is the first, and to date, the only, contemporaneous critical account of the FBI to be published by a mainstream publisher. He is also an independent investigative producer. He is the former editor and senior writer for the National Reporter, a publication specializing in reporting on the CIA.

According to the Central Intelligence Agency itself, as reported by the House Intelligence Committee, "The Clandestine Service of the CIA is the only part of the Intelligence Community, indeed of government, where hundreds of employees on a daily basis are directed to break extremely serious laws in countries around the world. A safe estimate is that several hundred times every day (easily 100,000 times a year), officers engage in highly illegal activities." (pp. 115, 116)

The national security of the United States requires that more than 100,000 extremely serious crimes be committed every year. The [House Intelligence] Committee expressed no legal or ethical concerns about these crimes. The committee indicated that it did not matter that laws were broken because they were laws of other countries. The CIA [is] committing crimes against humanity with de facto impunity and Congressional sanctioning. (pp. 116, 117)

Government documents, including CIA reports, show that the CIA's crimes include terrorism, assassination, torture, and systematic violations of human rights. The documents show that these crimes are part and parcel of deliberate CIA policy. The report notes that CIA personnel are "directed" to commit crimes. (p. 117)

CIA documents show that the CIA created, trained, and armed death squads in Guatemala as part of its coup and destabilization of the democratically elected government in 1954. In Honduras, the CIA's own inspector general reported that paid CIA assets at the highest level created and ran a death squad which, according to the Honduran government, murdered at least 184 people. The House Intelligence Committee's only concern regarding these brutal CIA informants and other CIA offenders was that they might be arrested and prosecuted. The committee did not advise the CIA to cease or limit its lawlessness. The Senate Intelligence Committee proposed a bill that would immunize CIA offenders who violate treaties and international agreements while following orders. The bill passed both houses of Congress and was signed into law by President Bill Clinton on December 27, 2000. (pp. 117-118)

[This law] means that the Constitution does not apply to the CIA or any US intelligence personnel. Why? Because the constitution provides that all treaties are the supreme law of the land. Not just law, but the supreme law - no exceptions. There was not a peep from the mass media about any of this even though such a story would not have affected corporate sponsorship or profits. (pp. 119)

The intelligence committees recommended that the "aggressive recruitment" of "terrorist informants who have human rights violations in their background" be "one of the highest priorities." Within months of instituting the guidelines, incoming CIA director George Tenet assured Congress that not a single unsavory applicant had been rejected. (pp. 120, 121)

Former ambassador Robert White wrote that Manuel Noriega of Panama, Colonel Julio Alpirez of Guatemala, General Gustavo Alvarez Martinez of Honduras, Colonel Nicolas Carranza of El Salvador, and Emmanuel Constant of Haiti, all major human rights abusers, were CIA informants who "enjoyed profitable contractual arrangements with the CIA not because they were particularly important sources of information, but because they served as paid agents of influence who promoted actions or policies favored by the CIA in that country." (p. 122)

Former CIA General Counsel Sporkin revealed that the CIA, not the president, creates findings to fit preordained covert operations and sends the findings to the president for his signature. (p. 126, 127)

There is next to no meaningful coverage ever of the CIA in the mainstream media, let alone analysis. The few exceptions prove the rule. In 1984, I was involved in one such exception. ABC hired me to help produce a story about an investment firm in Hawaii that was heavily involved with the CIA. I had earlier provided the same story to BBC's Newsnight, which aired it. The story was fully documented, and nobody, including the CIA, was able to disprove the charges. Part of the report charged that the CIA had plotted to assassinate an American, Ron Rewald, the president of [the investment firm]. The ABC report provoked a brutal response from the CIA. The CIA demanded a full retraction without providing any counterproof other than their denial. (pp. 130, 131)

At the center of the uproar was Scott Barnes who said on camera that the CIA had asked him to kill Rewald. After the show aired, CIA officials met with ABC News executive David Burke. They presented no evidence to counter the charges made in the program. Nonetheless, Burke was sufficiently impressed "by the vigor with which they made their case" to order an on-air "clarification" in which Peter Jennings acknowledged the CIA's position but stood by the story. But that was not good enough. [CIA Director William] Casey called ABC Chairman Leonard H. Goldenson. The call led to three meetings between ABC officials and Stanley Sporkin, CIA general counsel. On November 21, 1984, despite all the documented evidence presented in the program, Peter Jennings reported that ABC could no longer substantiate the charges, and that "We have no reason to doubt the CIA's denial." He presented no evidence supporting the CIA's position. (pp. 131, 132)

That same day, the CIA filed a formal complaint with the FCC, written by Sporkin and signed by [CIA Director] Casey, charging that ABC had "deliberately distorted" the news. Casey asked that ABC be stripped of its TV and radio licenses. This was the first time in the history of the country that a government agency had formally attacked the press. Yet, there was no uproar. (p. 132)

During this time, Capital Cities Communications was maneuvering to buy ABC. [CIA Director] Casey was one of the founders of Cap Cities. Cap Cities bought ABC for $3.5 billion, which was called a "bargain rate" by the trade media. Besides Casey, two other founders of Cap Cities had extensive ties to the intelligence community. Within months, the entire investigative unit [of ABC] was dispersed, and the commentator on the Rewald program was assigned to covering beauty pageants. Needless to say, my contract was not renewed. (pp. 132, 122)

For Mr. Kelly's book Tainting Evidence: Inside the Scandals at the FBI Crime Lab, see amazon.com.

Carl Jensen, Ph.D., founder and director emeritus of Project Censored, America's longest running research project on mass media censorship, has been involved with the media for more than 50 years as a daily newspaper reporter, weekly newspaper publisher, public relations practitioner, advertising executive, educator, and author. Jensen is author of the 1990-1996 annual Project Censored yearbooks, Censored: The News That Didn't Make the News ... and Why. He has won numerous awards for his work.

There were 50 major media corporations in 1993, and now there are only about half a dozen. Corporate socialization has been exacerbated by the multibillion-dollar mergers that created international giants such as AOL Time Warner, Disney, General Electric, News Corporation, and Viacom. (pp. 425-428)

Shortly after the outbreak of the First Terrorist War of the 21st Century, I was reminded of what US Senator Hiram Johnson said during World War I: "The first casualty when war comes, is truth." Post-September 11, 2001, the free flow of information in America is slowing to a carefully monitored trickle. The president of the US says he can only trust eight members of Congress. The attorney general admonishes Congress to pass the controversial Anti-Terrorism Act without debate. The national security adviser cautions TV networks not to broadcast press conferences with Taliban leaders because they may contain hidden messages. The military tells the press this is a "different war" and thus it can't observe the 1992 agreement allowing the media more access to information. The president's press secretary warns the media and all Americans to watch what they say and watch what they do. These are ominous signs for democracy. (pp. 432, 433)

In the same way that we survived Pearl Harbor, we will survive the Sept. 11 terrorist attack. In the meantime, let us not be terrorized into giving up any of our constitutionally guaranteed rights. (p. 434)

Robert McChesney has written or edited seven books and is currently research professor at the Institute of Communications Research at the University of Illinois. He has made more than 500 radio and TV appearances and has been the subject of nearly 50 published interviews.

Professional journalism had three distinct biases built into it, biases that remain to this day. First, it regarded anything done by official sources, for example, government officials and prominent public figures, as the basis for legitimate news. Second, professional journalism posited that there had to be a news hook or a news peg to justify a news story. [This] helped to stimulate the birth and rapid rise of the public relations (PR) industry. Surveys show that PR accounts for anywhere from 40 to 70 percent of what appears as news. The third bias is that [professional journalism] smuggles in values conducive to the commercial aims of the owners and advertisers as well as the political aims of the owning class. The affairs of government are subjected to much closer scrutiny than the affairs of big business. The genius of professionalism in journalism is that it tends to make journalists oblivious to the compromises with authority they routinely make. (pp. 440, 441)

Professional journalism equates the spread of "free markets" with the spread of democracy. To the US elite, however, democracy tends to be defined by their ability to maximize profit in a nation, and that is, in effect, the standard of professional journalism. (p. 442)

[There] is the striking consolidation of the mass media from hundreds of significant firms to an integrated industry dominated by less than ten enormous transnational conglomerates and rounded out by no more than another fifteen very large firms. The first tier giants include AOL Time Warner, Disney, Viacom, News Corporation, Bertelsmann, Vivendi Universal, Sony, AT&T, and General Electric. The nine or ten largest media conglomerates now almost all rank among the 300 largest firms in the world; in 1965, there were barely any media firms among the five hundred largest companies in the world. (p. 444)

The largest ten media firms own all the US TV networks, most of the TV stations in the largest markets, all major film studios, all major music companies, nearly all of the cable TV channels, much of the book and magazine publishing [industry], and much, much more. The logic of mass media industries is that a firm can no longer compete if it is not part of a larger conglomerate. General Electric's NBC is the only commercial TV network that does not own a major Hollywood film studio. (pp. 444, 445)

Expensive investigative journalism - especially that which goes after powerful corporate or national security interests – is discouraged. Largely irrelevant human interest/tragedy stories get the green light for extensive coverage. These are cheap, easy to cover, and they never antagonize those in power. The mass media companies claim they are responding to demand. (p. 445)

Throughout the 1980s and 1990s, real income declined or was stagnant for the lower 60 percent, while wealth and income for the rich skyrocketed. By 1998, discounting home ownership, the top 10 percent of the population claimed 76 percent of the nation's net worth. More than half is accounted for by the richest 1 percent. The Washington Post has gone so far as to describe ours as a nearly "perfect economy," which [reveals] the vantage point of the corporate news media. And it does appear more and more perfect the higher one goes up the socioeconomic ladder. (pp. 447, 448)

The rate of incarceration has more than doubled since the late 1980s. The US now has five times more prisoners per capita than Canada and seven times more than the whole of Western Europe. The US has 5 percent of the world's population and 25 percent of the world's prisoners. Nearly 90 percent of prisoners are jailed for nonviolent offenses, often casualties of the so-called drug war. It is a debate among Democrats and Republicans over who can be "tougher" on crime, hire more police, and build more prisons. Almost overnight, the prison-industrial complex has become a big business and a powerful lobby for public funds. (p. 448)

In the year 2000, a Texas man received 16 years in prison for stealing a Snickers candy bar, while four executives at Hoffman-LaRoche were found guilty of conspiring to suppress and eliminate competition in the vitamin industry in what the Justice Department called perhaps the largest criminal antitrust conspiracy in history. The four executives were fined anywhere from $75,000 to $350,000. They received prison terms ranging from three all the way up to four months. (p. 449)

The propagandistic nature of the war coverage was made crystal clear by AOL Time Warner's CNN a few weeks after the war began in Afghanistan. CNN president Walter Isaacson authorized CNN to provide two different versions of the war: a more critical one for the global audience and a sugarcoated one for Americans. Isaacson instructed the domestic CNN to be certain that any story that might undermine support for the US war be balanced with a reminder that the war on terrorism is a response to the heinous attacks of September 11. (p. 452)

We need to press for the overhaul of the media system, so that it serves democratic values rather than the interests of capital. The US media system has nothing to do with the wishes of the Founding Fathers and even less to do with the workings of some alleged free market. To the contrary, the media system is the result of laws, government subsidies, and regulations made in the public's name, but made corruptly behind closed doors without the public's informed consent. The largest media firms are all built on top of the profits generated by government gifts of monopoly rights. It is impossible to conceive of a better world with a media system that remains under the thumb of Wall Street and Madison Avenue, under the thumb of the owning class. It is nearly impossible to conceive of a better world without some changes in the media status quo. We have no time to waste. (p. 453)

Reference[iv] Article – Book – Into the Buzzsaw: Leading Journalists Expose the Myth of a Free Press – Edited by Christina Borjesson, Extracted from the Internet January 13, 2019 @ https://www.wanttoknow.info/massmedia

Clearing the Black Smoke of Fake News

A 1938 radio broadcast, "The War of the Worlds," dramatized an invasion by Martians whose technological powers vastly surpassed those of Earth's human defenders. At the program's climax, an "eyewitness" described the desperate scene as Martian machines attacked New York: "This is the end now... black smoke, drifting over the city. People in the streets see it now. They're running towards the East River... thousands of them, dropping in like rats. Now the smoke's spreading faster... People trying to run away from it, but it's no use."

October 30th marks the eightieth anniversary of the broadcast, which is legendary for having supposedly led to mass panic throughout the United States. Directed and narrated by Orson Welles, the most famous episode of CBS's "Mercury Theatre on the Air" is often invoked as a cautionary example of media's mighty influence and the public's vulnerability to media-induced hysteria. Revisiting the historic broadcast provides an opportunity to revise our shared understanding of what actually happened on the night of October 30, 1938, yielding insights relevant to contemporary concerns about "fake news."

Confronted by proprietary algorithms that covertly shape our news feeds, disingenuous pundits such as Alex Jones, and opportunistic political leaders including a president who consistently villainizes the press as "the enemy of the people," many of us may conclude that "it's no use" trying to find news worth trusting. Pervasive distrust of formerly respected information sources is paralleled by a glut of disinformation and propaganda that threatens, much like Martian black smoke, to suffocate our democracy.

The popular understanding that "War of the Worlds" caused public chaos is at odds with the facts. Media scholars Jefferson Pooley and Michael J. Socolow have identified two factors that explain enduring tall tales about the broadcast. First, it took place when radio, a newly-developing medium, had begun to compete successfully with print for audience attention and advertising revenues. Newspaper editors sensationalized the panic in hopes of undermining trust in radio and encouraging government regulation of it. Second, the publication in 1940 of Hadley Cantril's "The Invasion from Mars," one of the first empirical studies of media effects on audiences, helped to reinforce the broadcast's reputation as a prime example of mass hysteria.

Professional conflicts and financial motives led Cantril to overstate his study's findings.

Its most dramatic claims about "widespread panic" were not supported by the data, which was collected, and most carefully analyzed, by one of Cantril's research assistants, Herta Herzog. Herzog and a team of four other researchers (all female) interviewed 135 audience members, most of whom were chosen for the study because they had supposedly been frightened by the broadcast.

These interviews showed that, although many listeners found the broadcast "exciting," few of them actually believed the program was a live report of an alien invasion. Instead, Herzog found that many audience members had engaged in "checking up," the term she used to identify what listeners did to verify whether the broadcast was real or not.

In "War of the Worlds," after all human efforts to defeat the Martians had failed, what stopped the invaders was accidental exposure to bacteria "against which their systems were unprepared." In 1938, despite radio's growing influence and Orson Welles's canny sense of realistic drama, what prevented an outbreak of widespread panic was "critical ability," which wary audience members employed to determine that the extraordinary broadcast was fiction.

Much as human weapons proved useless against Martian invaders, we should not count on private corporations such as Facebook and Google to stop the spread of false information by blocking problematic content or offensive social media users. The enduring lesson of Herta Herzog's underappreciated research is that individuals engaging their critical abilities can forestall collective hysteria. We should support public institutions that strengthen our capacities for critical thinking—including schools, libraries, and media literacy programs—as our best defense against the black smoke of false news and misinformation.

Reference[v] Article - Clearing the Black Smoke of Fake News, October 28, 2018October 28, 2018. Written by Andy Lee Roth whom is the associate director of Project Censored and co-editor of the Project's newest book, Censored 2019: Fighting the Fake News Invasion. Extracted from the Internet January 13, 2019 @ https://www.projectcensored.org/clearing-the-black-smoke-of-fake-news/

Media Deception & Public Perception

We live in a dirty and dangerous world. There are some things the general public does not need to know and shouldn't. I believe democracy flourishes when the government can take legitimate steps to keep its secrets and when the press can decide whether to print what it knows. ~ Katharine Graham, former owner of the Washington Post

The first lesson extracted from the Hidden Knowledge Course explores the ways in which we are collectively deceived and sometimes even deceive ourselves. In order to make a difference in the world, we can be most effective when we understand the basic inner workings of the all that is hidden and then come together to transform the shadows both within ourselves and in the world.

This shadow exploration starts here with media deception and the media's role in influencing public perception. First, please watch a fascinating and revealing six-minute video clip from Canada's Vision TV on a little-known U.S. government trial in 1999 involving the Martin Luther King, Jr. assassination. https://www.personalgrowthcourses.net/video/martin_luther_king_jr_mlk_assassination_video

Why is it that so few people are aware of this amazing trial and its verdict? Is the media somehow controlled in reporting on certain topics?

Next is an engaging essay which presents concise summaries of revealing writings by a number of award-winning, courageous journalists further suggesting that certain topics are censored from the news by media ownership. These journalists reported on powerful, disturbing stories at the risk of losing their jobs in order to bring highly important news to the public.

While you read some of the eye-opening information in section following titled Leading Journalists Expose Major News Cover-ups by Media Corporations, remember that by working together, we can change all of this.

Many of you may already know or have suspected how corporate ownership of the media has tended to promote fear and sensationalism while neglecting some of the most important issues our world is facing. Others may feel a bit overwhelmed after reading the revealing stories of these courageous, award-winning journalists.

For those who were not aware of media complicity in keeping certain key information from the public, please realize that many who have taken this course were shocked and disturbed when they first learned about some of these things. Yet we also know without doubt that every one of us can make a difference and together we can be the change and build the world we envision.

Many people would rather not know about the major manipulations happening in our world. Believing that it is better to focus only on the positive, they choose to avoid looking into the shadows. Yet would you ignore a friend who was about to harm someone or commit suicide?

As long as we collectively choose ignorance over awareness, these manipulations will continue. In fact, they will likely grow in magnitude until people are finally forced to open their eyes and deal with the consequences.

The sooner each of us decides that we do want to know of important matters that are being hidden, and that we are willing to invite others to open their eyes, the more easily we will be able to transform our world in a direction that supports the good of all who share our world with us.

This book includes powerful information on the shadow side of our world, the purpose is always one of inspiring you individually and all of us collectively to make a difference. Consider that this can be a tremendous opportunity for us to build a brighter future for our world.

Never assume the obvious is true. ~ William Safire

Ours is not the task of fixing the entire world, but of stretching out to mend the part of the world that is within our reach. ~ Clarissa Pinkola Estes

Project Censored

One organization stands out in doing powerful work to educate the public on media deception. Project Censored is an inspiring university program dedicated to reporting major news stories which the mainstream media either largely or completely failed to cover. Please read the revealing information by Project Censored at the following link. www.WantToKnow.info/mass_media/censored/censored-news-stories-18

Thanks to the amazing tool of the Internet, those who want to be informed of important matters neglected by the major media can find such information on reliable websites such as like projectcensored.org and WantToKnow.info, which are designed specifically for this purpose. Certainly not all news on the Internet is reliable, but by using discernment, it is possible to find many quality websites (listed at the end of this article) with reliable reports of great importance which otherwise, would not be available.

Finding suppressed information on key people or events has become tremendously easier than it was just a few decades ago thanks to the empowering capabilities of the Internet which incidentally, was originally created by the military – more info here at https://en.wikipedia.org/wiki/DARPA).

The World Wide Web is most certainly a powerful tool of change. Without it, you would not be reading a lot of data in this book and the Hidden Knowledge Course would not be possible as well. The Internet is providing a giant leap towards greater openness and transparency on both an individual and global basis.

Power of Nightmares - A BBC Documentary

For the next section of this lesson, please read the web page and watch the third part of the BBC documentary Power of Nightmares at the link below. Each section of this three-part BBC series is 60 minutes. If you have time, all three parts have excellent information, though part three is enough for the purpose of this lesson. This documentary will give further context for the role of the media and politicians in the unfolding drama of our world. Links to written transcripts are included.
www.personalgrowthcourses.net/stories/power_of_nightmares

As is evident in the Power of Nightmares, many of the political and economic elite of this world use fear and "national security" as tools to maintain and increase their hold on power. By keeping the public in a state of fear and insecurity, it is much easier for those with hidden agendas to promote war and to channel more of our tax money to support the defense industry and the mega-corporations who profit handsomely from war.

With its increasing reliance on sensationalism to garner profits, the media all too often latches on to government propaganda laden with fear and promotes it widely – often on the front pages and in top headlines. When the evidence later shows that the fear-inducing news was greatly exaggerated or even completely false, this "correction" is rarely given much attention, so that the original perception of fear remains embedded in the public mind.

Change is in the Air

The extract from the Hidden Knowledge Course starts with the media, as it is one of the key ways in which vital information is kept hidden from the public. Though some reporters and editors truly want to bring all important information to the public, many among the elite ownership of the world's media have their own agenda which sometimes blocks key information from reaching the public. You will see more of this in your information awareness documented here.

Yet as humanity reaches a point of wanting to know more of what is going on, the Internet is now filling the role at which the old media is failing. As mentioned, numerous alternative media websites and even transformational sources are now available thanks to the interconnectivity of the World Wide Web. They are filling in the gap, educating the public and calling people to action. Change shows up in many unexpected ways, the Internet being just one of these.

Even as we expose this important information, let us examine how we at times also engage in similar behavior. Don't most of us occasionally broadcast each other's life dramas and traumas rather than work together to transform them? And how often do we choose to keep some of our own or our loved ones' deepest, darkest secrets from ever being aired?

Don't most of us have "skeletons in the closet," personal or family secrets that we are careful never to reveal, believing that this is what's best for all concerned? Let us commit to work towards supporting and broadcasting the very best in each other and towards ever greater transparency in all aspects of our lives.

Remember, too, that even those who own and control the media have a heart. They once were small children who knew little but love and laughter. As a way to see this, please watch this heart-warming, two-minute video clip of a baby laughing, and imagine that even those who would hurt or control us were once like this fun baby.
www.personalgrowthcourses.net/video/baby_laughing - Laughing Baby

And as the potent, one-minute video clip below demonstrates, the process of change is open to all.
www.personalgrowthcourses.net/video/metaphormosis - METAPHORmosis

Thanks for your willingness to explore the shadow material in this book. Again, the once Hidden Knowledge is not designed to convince you of anything. Yet it does present information that can expand your horizons and allow all of us to be more effective in making a difference in our lives and world. By learning to dance with the shadow and inviting the dark places to come to light, we are making a difference in our lives and world in a big way.

Uncensored News Websites and Lists

The WantToKnow.info email list specializes in sending out revealing news and powerful, little-known information which the major media fails to cover. On average one email is sent out every few days. Key media articles which were buried and should have been front page news are summarized with links available for those who want to read the full original article. Reliability and verifiability are top priority. If you want to be well informed and better able to deal with the challenges of today's world, this is a great place to start.

Truth Out, http://www.truthout.org (Free)

Rating–Interest 9 out of 10, Reliability 9.5 out of 10

Truth Out provides excellent, reliable information on important news that is often not well covered by the mainstream media. Subscribe and get one email a day listing headlines and a brief description of the most important news of the day. You choose which stories you want to read by clicking on the links provided. Some information shows a liberal (or at least anti-Bush) bias, though overall, this news service reports important information for people from across the political spectrum. Consider turning off your TV news and reading this concise daily summary of important news you won't find in the mainstream media.

Information Clearing House, http://www.informationclearinghouse.info Free

Rating—Interest 8.0 out of 10, Reliability 8.5 out of 10

Information Clearing House is an independent media source which attempts to balance distorted perceptions provided by the commercial media. A daily list of news headlines with links to informative stories focuses on news which goes unreported and underreported, particularly in the U.S. media. Though this service focuses heavily on war casualties, coverage of other important, revealing news is generally of a high quality. Excellent, revealing videos are also provided for free viewing on the website. You can sign up for a daily email or simply visit the website. The website and news often reveals an anti-Bush administration sentiment.

whatreallyhappened.com, http://www.whatreallyhappened.com (Free)

Rating–Interest 9 out of 10, Reliability 7.5 out of 10

This popular, non-partisan website offers a continual stream of important news that the mainstream often fails to cover in any significant way. Though some of their sources are not well known by the public, they do pay attention to reliability and often have powerful, revealing news you cannot find in the mainstream media. We recommend reading their news with a discerning eye, yet in so doing, you will likely find vital news not being covered on other, more reliable new websites.

Reference Article - Media Deception & Public Perception, Transformation Index lesson plan one extracted from the Internet January 14, 2019 at https://www.transformationteam.net/ttcourse/ttcourse02a[vi]

FreeTheTech Finds 5,784 Inventions Suppressed by U.S. Government Secrecy Orders

FreeTheTech is an organization that is looking to spread awareness of the suppression and bring greater transparency by creating a citizen review panel to assess the classified patents.

As reported last month via PRNewswire, over 5,000 inventions are being suppressed via the U.S. Government's classified patent program, according to official data obtained under a recent Freedom of Information Act request by the Federation of American Scientists. This means literally thousands of potentially ground-breaking inventions and technologies are being withheld from the public.

A new project called FreeTheTech.org based in Austin, Texas is raising awareness of the issue through a variety of means, including its new website, a petition on Change.org, and a Billboard fundraising program. Their primary goal is simply to make the public aware that many potentially useful technologies are being suppressed, while their secondary goal is to audit the program via a citizen review panel.

What exactly is contained within the government's archive of 5,784 classified patents has been the source of rampant speculation for decades. However, a recent leak from the Category List of 1971 finally provides some real-world insight as categories from it include: Power Supply, Meteorology, Propulsion Systems, and Unique Materials & Devices. Interestingly, leaked data from the Power Supply section shows that modern-day solar-panel technology was being restricted some 50 years ago. As a result, one can only imagine what's being suppressed today.

The origin of the government's program is the Invention Secrecy Act of 1951, which evolved invention secrecy from a temporary wartime agenda into a permanent one. As a result, every time a patent is submitted by an inventor today, it can be flagged for review by a multitude of government agencies. All it takes is a single opinion by one of them to keep the invention from ever seeing the light of day.

Patents can be restricted not only for being a national security threat, but also for disrupting economic stability. This means that any number of potentially ground-breaking inventions that challenge the status quo may be restricted and suppressed. FreeTheTech.org's aim is to end all of the speculation by forming a citizen review panel consisting of both experts and general citizens alike. They would independently, yet privately, review all classified patents to yield greater transparency and accountability.

"While promoting our Campaign on Facebook," says project founder Benjamin Cohen, "we've had numerous users assume we're spreading false information due to their own lack of research into this subject matter. While we certainly understand their initial skepticism, the fact remains that this data comes directly from the government itself." Cohen points to the US Patent and Trademark Office's response letter to FOIA Request# F-18-00003, which reveals updated 2017 data showing the new figure of 5784 patents currently under secrecy orders.

Reference [vii] Article - FreeTheTech Finds 5,784 Inventions Suppressed by U.S. Government Secrecy Orders, written By Sam Davis whom is editor-in-chief of Power Electronics Technology magazine and website. Posted September 14, 2018. Extracted from the Internet January 10, 2019 @ https://www.powerelectronics.com/news/sorry-primary-kilogram-we-are-dumping-you . For more information, visit the project's main website at www.FreeTheTech.org.

Are You Overwhelmed or Upset With All The Craziness?

This message is for those who may be feeling upset, angry, or otherwise overwhelmed by all the craziness in our world and by any of the material on the WantToKnow.info website or within this book. The intention of presenting these materials is not at all to make anyone feel overwhelmed or upset.

The combined intention is to help people understand what has been hidden from the public by various forces, and to inspire all of us to join together in building a brighter future for us all. There is no doubt that everyone of us can make a difference in what is happening in our world.

There are clear reasons why all of this can be overwhelming. Most of us over the years have developed fairly well-established ideas of how democracy, government, and businesses function in our nation and world. Our educational system and the media generally reinforce beliefs we developed from an early age in the way the world works.

It is not easy to be open to evidence which seriously challenges these beliefs, even if the information is verifiable and from reliable sources. When presented with information which questions our basic understanding of the world, it's natural to feel uneasy and overwhelmed. Yet we encourage you not to despair. Though it can be overwhelming at times, choosing knowledge and understanding in the long run allows us to work together to build a brighter future.

It's only natural that some become upset or angry to discover how much they have been deceived. Many contributors who developed material posted on WantToKnow.info or within this book went through a period of shock and anger upon first learning about the major cover-ups before they eventually shifted to work for change in a positive way. Some individuals may need to go through a period of anger at being deceived. Though it may be important at times to give our anger a voice, in the long run, anger will not change our world for the better, and excessive anger will likely only lead to more of the same.

The underlying intention of providing all of these materials is to open you to the possibility that there is much more than you might have imagined going on behind the scenes in the world – both destructive and constructive.

There is potential for good and bad in each one of us. When we choose to tap into the potential for good, we each have the ability to make positive changes in our lives and world. As ever increasing numbers of us come together in the intention of positive change – without being in denial of all of the destructive forces – we help to shift the vision of humanity towards a more supportive, cooperative way of life.

By opening to understanding the deep difficult parts both within ourselves and out in the world, we can then choose to learn from these challenges.

Through tapping into our abundant inner potential, we can collectively transform the difficult places both in our own lives and in the world as we co-create a more caring, supportive way of being and living together.

Some of you do not need to go into depth with all of this information, as you feel called in your life to focus on other important matters. We fully support you in this. Yet even so, we recommend that you at least take the time to read the two-page summaries of each major category listed on WantToKnow.info, so that you have a basic understanding of what is happening and can thus make informed decisions. And don't miss their summaries of all the great reasons for hope and optimism.

Others will choose to dive deep into the mountains of reliable information provided and use it powerfully to work towards a brighter future. If you want to understand the deeper levels of all that is going on, we encourage you to find the time to read each major topic.

Whether you choose to just read the short summaries or to explore our information extensively, we encourage you to read and explore at a pace that is appropriate for you. If you find yourself having a difficult reaction, consider taking a break and possibly enjoying some of the excellent inspirational articles and resources provided on WantToKnow.info to keep things in perspective.

There are many wonderful things also happening in the world which are being hidden from public view. The Inspiration Center will introduce you to some of these exciting developments, as will the hope and optimism page previously mentioned.

And while exploring the information on cover-ups and corruption, we encourage you to be skeptical. Some of what you read may at first seem quite unbelievable. Do a little research using the reliable sources provided and determine for yourself whether there is truth to the information provided. And we recommend opening to whatever form of spiritual guidance you might have in asking for discrimination, open-mindedness, understanding, and inspiration as you read this material in this book.

Many people don't want to know about the major cover-ups going on in our world. Feeling relatively secure and content in their personal lives, they choose to avoid matters which might disturb their contentment. Yet as long as we choose complacency over awareness, these major cover-ups will continue. In fact, they will likely grow in magnitude until people are finally forced to open their eyes and deal with the consequences.

The sooner each of us decides that we do want to know, and that we are willing to invite others to open their eyes, the more easily we will be able to build a world that supports the good of all of us.

For those who are asking what we can do about all this, there are many ways we can work together to strengthen democracy and build a better world. For sure, by sharing this information with your friends and colleagues, you can play a key role in restoring a true democracy of the people, by the people, and for the people.

Whenever you find yourself slipping into fear or anger with this material, remember that we now have an opportunity to do something about it. Please email your friends and colleagues encouraging them to explore this book, WantToKnow.info and the many links provided. By making good use of the Internet to spread this knowledge around the globe, you can play a vital role in building the critical mass necessary to bring about positive, empowering changes in our lives and in our world.

The more we share this information with others, the more chance there is of transforming our world into a safer, healthier place to live.

Reference [viii]Article - Are You Overwhelmed or Upset With All The Craziness?. Posted on website WantToKnow.info which is an educational website. Extracted from the Internet by permission January 12, 2019 @ https://www.wanttoknow.info/overwhelmed

Section Three

Free Energy Turmoil

Free energy suppression conspiracy theory

Free energy suppression (or new energy suppression) is a conspiracy theory that technologically viable, pollution-free, no-cost energy sources are being suppressed by government, corporations, or advocacy groups. Devices allegedly suppressed include perpetual motion machines, cold fusion generators, torus-based generators, reverse-engineered extraterrestrial technology and other generally unproven, low-cost energy sources.

The alleged suppression (or weakening) is claimed to have occurred since the mid-19th century and allegedly perpetrated by various government agencies, corporate powers, special interest groups, and fraudulent inventors. The special interest groups are usually claimed to be associated with the fossil fuel or nuclear industry, whose business model would be threatened.

Claims of suppression include:

The claim that the scientific community has controlled and suppressed research into alternative avenues of energy generation via the institutions of peer review and academic pressure.

The claim that devices exist which are capable of extracting significant and usable power from pre-existing unconventional energy reservoirs, such as the quantum vacuum zero point energy, for little or no cost, but are being suppressed

Some examples of individuals who were allegedly suppressed, harassed or killed for their research are:

Thomas Henry Moray - (August 28, 1892 - May 18, 1974) was an inventor from Salt Lake City, Utah. "In the 1920s T. Henry Moray invented a 'free energy' device that reportedly output 50 kilowatts of electricity. It could not be explained by standard science at that time. The electricity exhibited a strange 'cold current' characteristic where thin wires could conduct appreciable power without heating. Moray suffered ruthless suppression, and eventually the device was destroyed. Despite it all, he received a US patent 2,460,707 in February 1949, after a process of 17 years in discussions with the patent office. The title of the patent is "Electrotherapeutic Device", and although radiotherapy is mentioned, no details are given.

Moray's invention was not commercially successful because of the high manufacturing cost compared to the amount of power produced and the small market for atomic powered batteries. Many unsubstantiated claims have been made in connection with attempts to sell books or ask for money. One claim was five kilowatts of electricity produced from a device costing sixty thousand dollars to build in 1926. If true, it represents a high cost for electricity compared to other sources, except in special situations like space research

A counter culture has developed with claims about alternative energy, citing Moray as a leading example of lost opportunity and of free energy suppression. Since Moray patented his invention with detailed drawings and further described his ideas in books he wrote, the economics and technical operation can be understood with conventional science and engineering. A substantial reduction of manufacturing cost would be required to make wider use of power supplies based on Moray's invention.

Stanley Meyer – Patented the water fuel cell that purportedly split water into its component elements, hydrogen and oxygen. The hydrogen gas was then burned to generate energy, a process that reconstituted the water molecules. According to Meyer, the device required less energy to perform electrolysis than the minimum energy requirement predicted or measured by conventional science. The mechanism of action was alleged to involve "Brown's gas", a mixture of oxyhydrogen with a ratio of 2:1, the same composition as liquid water; which would then be mixed with ambient air (nitrogen, oxygen, carbon dioxide, carbon monoxide, methane, chlorofluorocarbons, free radicals/electrons, radiation, among others). If the device worked as specified, it would violate both the first and second laws of thermodynamics, allowing operation as a perpetual motion machine.

Throughout his patents Meyer used the terms "fuel cell" or "water fuel cell" to refer to the portion of his device in which electricity is passed through water to produce hydrogen and oxygen. Meyer's use of the term in this sense is contrary to its usual meaning in science and engineering, in which such cells are conventionally called "electrolytic cells". Furthermore, the term "fuel cell" is usually reserved for cells that produce electricity from a chemical redox reaction, whereas Meyer's fuel cell consumed electricity, as shown in his patents and in the circuit pictured on the right. Meyer describes in a 1990 patent the use of a "water fuel cell assembly" and portrays some images of his "fuel cell water capacitor". According to the patent, in this case "... the term 'fuel cell' refers to a single unit of the invention comprising a water capacitor cell ... that produces the fuel gas in accordance with the method of the invention."

In 1996 Meyer was sued by two investors to whom he had sold dealerships, offering the right to do business in Water Fuel Cell technology. His car was due to be examined by the expert witness Michael Laughton, Professor of Electrical Engineering at Queen Mary, University of London and Fellow of the Royal Academy of Engineering. However, Meyer made what Professor Laughton considered a "lame excuse" on the days of examination and did not allow the test to proceed. According to Meyer, the technology was patent pending and under investigation by the patent office, the Department of Energy and the military. His "water fuel cell" was later examined by three expert witnesses (names unknown) in court who found that there "was nothing revolutionary about the cell at all and that it was simply using conventional electrolysis." The court found Meyer had committed "gross and egregious fraud" and ordered him to repay the two investors their $25,000

Stanley Meyer died suddenly on March 20, 1998, while dining at a restaurant. His brother claimed that during a meeting with two Belgian investors in a restaurant, Meyer suddenly ran outside, saying "They poisoned me". After an investigation, the Grove City police went with the Franklin County coroner report that ruled that Meyer, who had high blood pressure, died of a cerebral aneurysm. Stanley Meyer's bizarre death at age 57 ended work that, if proved valid, could have ended reliance on fossil fuels. Some of Meyer's supporters believe that he was assassinated to suppress his invention.

Eugene Mallove - Eugene Franklin Mallove (June 9, 1947 – May 14, 2004) was an American scientist, science writer, editor, and publisher of Infinite Energy magazine, and founder of the nonprofit organization New Energy Foundation. He was a proponent of cold fusion, and a supporter of its research and related exploratory alternative energy topics, several of which are sometimes characterized as "fringe science".

Mallove authored Fire from Ice, a book detailing the 1989 report of tabletop cold fusion from Stanley Pons and Martin Fleischmann at the University of Utah. Among other things, the book claims the team did produce "greater-than-unity" output energy in an experiment successfully replicated on several occasions, but that the results were suppressed through an organized campaign of ridicule from mainstream physicists, including those studying controlled thermonuclear fusion, trying to protect their research and funding.

Mallove was murdered in 2004 while cleaning out his former childhood home, which had been rented out. Three people have been arrested and charged in connection with the killing; two were convicted of first-degree manslaughter and murder, in 2012 and 2014; the third pleaded guilty to obstruction of justice in 2015.

Reference[ix] Article - Free energy suppression conspiracy theory Extracted from the Internet January 15, 2019 @ https://en.wikipedia.org/wiki/Free_energy_suppression_conspiracy_theory#cite_note-16

Free Energy?

Conventional Energy Producers Would Go Out of Business

Ninety to a hundred years ago, everybody "knew" that a heavier-than-air machine could not possibly fly. It would violate the "laws" of physics. All of the "experts" and "authorities" said so.

For example, Simon Newcomb declared in 1901: "The demonstration that no possible combination of known substances, known forms of machinery and known forms of force, can be united in a practical machine by which man shall fly long distances through the air, seems to the writer as complete as it is possible for the demonstration of any physical fact to be."

Fortunately, a few SMART people such as the Wright Brothers did NOT accept such pronouncements as the final word. Now we take airplanes for granted, (except when they crash).

Today, orthodox physicists and other "scientists" are saying similar things against several kinds of 'Free Energy' Technologies, using negative terms such as "pseudo-science" and "perpetual motion", and citing so-called "laws" which assert that "energy cannot be created or destroyed" ("1st law of thermodynamics") and "there is always a decrease in useful energy" ("2nd law of thermodynamics"). The physicists do not know how to do certain things, so they ARROGANTLY declare that those things cannot be done. Such PRINCIPLES OF IMPOTENCE are COMMON in orthodox modern "science" and help to cover up INCONSISTENCIES and CONTRADICTIONS in orthodox modern theories.

Free Energy Inventions are devices which can tap a seemingly UNLIMITED supply of energy from the universe, with- OUT burning any kind of fuel, making them the PERFECT SOLUTION to the world-wide energy crisis and its associated pollution, degradation, and depletion of the environment.

Most Free Energy Devices probably do not create energy, but rather tap into EXISTING natural energy sources by various forms of induction. UNLIKE solar or wind devices, they need little or no energy storage capacity, because they can tap as much energy as needed WHEN needed. Solar energy has the DIS-advantage that the sun is often blocked by clouds, trees, buildings, or the earth itself, or is reduced by haze or smog or by thick atmosphere at low altitudes and high latitudes. Likewise, wind speed is WIDELY VARIABLE and often non-existent. Neither solar nor wind power are suitable to directly power cars and airplanes. Properly designed Free Energy Devices do NOT have such limitations.

For example, at least three U.S. patents (#3,811,058, #3,879,622, and #4,151,431) have so far been awarded for motors that run EXCLUSIVELY on permanent MAGNETS, seemingly tapping into energy circulating through the Earth's magnetic field. The first two require a feedback network in order to be self-running. The third one, (as described in detail in "Science & Mechanics" magazine, Spring 1980), requires critical sizes, shapes, orientations, and spacings of magnets, but NO feedback. Such a motor could drive an electric generator or reversible heat pump in one's home, YEAR ROUND, FOR FREE. [Complete descriptive copies of U.S. patents are $3.00 each from the U.S. Patent Office, 2021 Jefferson Davis Hwy., Arlington, VA 22202; correct 7-digit patent number required. Or try getting copies via your local public or university library's inter-library loan dept..]

A second type of free-energy device, such as the 'Gray Motor' (U.S. Patent #3,890,548), the 'Tesla Coil', and the unpatented motor of inventor Joseph Newman, taps ELECTRO- MAGNETIC energy by INDUCTION from 'EARTH RESONANCE' (about 12 cycles per second plus harmonics). They typically have a 'SPARK GAP' in the circuit which serves to SYNCHRONIZE the energy in the coils with the energy being tapped. It is important that the total 'inductance' and 'capacitance' of the Device combine to 'RESONATE' at the same frequency as 'EARTH RESONANCE' in order to maximize the power output. This output can also be increased by centering the SPARK GAP at the 'NEUTRAL CENTER' of a strong U-shaped permanent magnet. In the case of a Tesla Coil, slipping a 'TOROID CHOKE COIL' around the secondary coil will enhance output power. ["Earth Energy: Fuelless Propulsion & Power Systems", by John Bigelow, 1976, Health Research, P.O. Box 70, Mokelumne Hill, CA 95245.]

During the 1930's, an Austrian civil engineer named Viktor Schauberger invented and partially developed an 'IMPLOSION TURBINE' (German name, 'ZOKWENDLE'), after analyzing erosion, and lack of erosion, in differently shaped waterways, and developing sophisticated mathematical equations to explain it. As described in the book "A Breakthrough to New Free-Energy Sources", by Dan A. Davidson, 1977, water is pumped by an IMPELLER pump through a LOGARITHMIC-SPIRAL-shaped coil of tubing until it reaches a CRITICAL VELOCITY. The water then IMPLODES, no longer touching the inside walls of the tubing, and drives the pump, which then converts the pump's motor into an ELECTRIC GENERATOR. The device seems to be tapping energy from that of the earth's rotation, via the 'Coriolis effect', LIKE A TORNADO.

A fourth type of Free Energy Device is the 'McClintock Air Motor' (U.S. Patent #2,982,261), which is a cross between a diesel engine (it has three cylinders with a compression ratio of 27 to 1) and a rotary engine (with solar and planetary gears). It burns NO FUEL, but becomes self-running by driving its own air compressor. This engine also generates a lot of heat, which could be used to heat buildings; and its very HIGH TORQUE makes it ideal for large trucks, preventing their slowing down when climbing hills. [David McClintock is also the REAL original Inventor of the automatic transmission, differential, and 4-wheel drive.]

Crystals may someday be used to supply energy, as shown in the Star Trek shows, perhaps by inserting each one between metal capacitor plates and bombarding it with a beam of particles from a small radioactive source like that used in a common household smoke detector.

One other energy source should be mentioned here, despite the fact that it does not fit the definition of Free Energy. A Bulgarian-born American Physicist named Joseph Maglich has invented and partially developed an atomic FUSION reactor which he calls 'Migma', which uses NON-radioactive deuterium as a fuel [available in nearly UNLIMITED quantities from sea water], does NOT produce radioactive waste, can be converted DIRECTLY into electricity (with-OUT energy-wasting steam turbines), and can be constructed small enough to power a house or large enough to power a city. And UNLIKE the "Tokamaks" and laser fusion MONSTROSITIES that we read about, Migma WORKS, already producing at least three watts of power for every watt put in. ["New Times" (U.S. version), 6-26-78, pages 32-40.]

And then there are the 'cold fusion' experiments that have been in the news lately, originally conducted by University of Utah researchers B. Stanley Pons and Martin Fleischmann. Some U.S. Navy researchers at the China Lake Naval Weapons Center in California, under the direction of chemist Melvin Miles, finally took the trouble to collect the bubbles coming from such an apparatus, had them analyzed with mass-spectrometry techniques, and found HELIUM 4, which PROVES that atomic FUSION did indeed take place, and enough of it to explain the excess heat generated.

There are GOOD INDICATIONS that the two so-called "laws" of thermodynamics are NOT so "absolute". For example, the late Physicist Dewey B. Larson developed a comprehensive GENERAL UNIFIED Theory of the physical universe, which he calls the 'Reciprocal System', (which he describes in detail in several books such as "Nothing But Motion" (1979) and "The Universe of Motion" (1984)), in which the physical universe has TWO DISTINCT HALVES, the material half and an anti-matter half, with a

CONTINUOUS CYCLE of matter and energy passing between them, with-OUT the "heat death" predicted by thermodynamic "laws". His Theory explains the universe MUCH BETTER than modern orthodox theories, including phenomena that orthodox physicists and astronomers are still scratching their heads about, and is SELF-CONSISTENT in every way. Some Free Energy Devices might be tapping into that energy flow, seemingly converting "low-quality energy" into "high-quality energy".

Also, certain religious organizations such as 'Sant Mat' and 'Eckankar' teach their Members that the physical universe is only the LOWEST of at least a DOZEN major levels of existence, like parallel universes, or analogous to TV channels, as described in books like "The Path of the Masters", by Julian Johnson, 1939, and "Eckankar: The Key to Secret Worlds", by Sri Paul Twitchell, 1969. For example, the next level up from the physical universe is commonly called the 'Astral Plane'. Long-time Members of these groups have learned to 'Soul Travel' into these higher worlds and report on conditions there. It seems plausible that energy could flow down from these higher levels into the physical universe, or be created at the boundary between them, given the right configuration of matter to channel it. This is supported by many successful laboratory-controlled experiments in PSYCHO-KINESIS throughout the world, such as those described in the book "Psychic Discoveries Behind the Iron Curtain".

In terms of economics, the market has FAILED. Inventors do not have enough money and other resources to fully develop and mass-produce Free Energy Equipment, and the conventional energy producer$ have no desire to do so because of their VE$TED INTERE$T$. The government is needed to intervene. If the government does not intervene, then the total supply of energy resources from the earth will continue to decline and will soon run out, prices for energy will increase, and pollution and its harmful effects (including the 'GREENHOUSE EFFECT', acid rain, smog, radioactive contamination, oil spills, rape of the land by strip mining, etc.) will continue to increase.

The government should SUBSIDIZE research and development of Free Energy by Inventors and universities, subsidize private production (until the producers can make it on their own), and subsidize consumption by low-income consumers of Free Energy Hardware.

The long-range effects of such government intervention would be wide-spread and profound. The quantity of energy demanded from conventional energy producer$ (coal mining companies$, oil companies$ and countries, electric utilities$, etc.) would drop to near zero, forcing their employees to seek work elsewhere. Energy resources (coal, uranium, oil, and gas) would be left in the ground. Prices for conventional energy supplies would also drop to near zero, while the price of Free Energy Equipment would start out high but drop as supply increases (as happened with VCR's, personal computers, etc.). Costs of producing products that require large quantities of energy to produce would decrease, along with their prices to consumers. Consumers would be able to realize the "opportunity costs" of paying electric utility bills or buying home heating fuel. Tourism would benefit and increase because travelers would not have to spend their money for gasoline for their cars. Government tax revenue from gasoline and other fuels would have to be obtained in some other way. And energy could no longer be used as a MOTIVE OR EXCUSE FOR MAKING WAR.

Many conventional energy producers would go out of business, but society as a whole and the earth's environment and ecosystems, would benefit greatly. It is the People, that government should serve, rather than the big corporation$ and bank$.

UN-altered REPRODUCTION and DISSEMINATION of this IMPORTANT Information is ENCOURAGED.

Reference[x] Article - Free Energy Technology Written by Robert E. Mc Elwaine, B.S., Physics, UW-EC Extracted from the Internet January 8, 2019@ https://www.bibliotecapleyades.net/ciencia/secret_projects/project119.htm

Scientists Claim To Tap The Free Energy Of Space

Subtitle: A promising new alternative energy source, neglected in the U.S. advances in the Far East

Physicist Bruce DePalma has a 100 kilowatt generator, which he invented, sitting in his garage. It could power his whole house, but if he turns it on, the government may confiscate it.

Harvard educated DePalma, who taught physics at the Massachusetts Institute of Technology for 15 years, claims that his electrical generator can provide cheap, inexhaustible, self-sustaining and nonpolluting source of energy, using principles that flout conventional physics and are still not fully understood. His N machine, as it is called, is said to release the "free energy" latent in the space all around us. DePalma views his device as an innovation that could help to end the world's dangerous dependence on finite supplies of oil, gas, and other polluting fossil based fuels.

Deceptive Simplicity:

The DePalma generator is essentially a simple magnetized flywheel, i.e. a magnetized cylindrical conductor rotating at high speed with the help of a motor. His astonishing claim is that the present versions of the N machine can generate up to five times more power than it consumes. This, of course, defies the basic law of the conservation of energy, which says that the output of energy cannot be more than the input. Most physicists simply refuse to look at DePalma's findings and dismiss his theories out of hand.

Yet "proof of principle" for his invention was apparently provided when a large N machine, dubbed the Sunburst, was built in 1978 in Santa Barbara California. The Sunburst machine was independently tested by Dr. Robert Kincheloe, professor emeritus of electrical engineering at Stanford University. in his 1986 report (presented to the Society for Scientific Exploration, San Francisco, 6/21/86), Kincheloe noted that the drag of the rotating magnetized gyroscope is only 13 to 20 percent of a conventional generator operating at an ideal 100 percent efficiency, the DePalma N machine could produce electricity at around 500 percent efficiency.

In Kincheloe's cautious summary: "DePalma may have been right in that there is indeed a situation here whereby energy is being obtained from a previously unknow and unexplained source. This is a conclusion that most scientists and engineers would reject out of hand as being a violation of accepted laws of physics and if true has incredible implications".

"The jury is still out on the DePalma N machine," says physicist Harold Puthoff, a senior fellow at the Institute for Advanced Studies in Austin, Texas. "It isn't clear where the reported excess energy is coming from--wheatear out of the electromagnetic field or as the result of some anomaly associated with rotating bodies in terms of inertia. The DePalma machine needs to be replicated on a broad scale to see if it actually works. Though I'm rather skeptical, I certainly would encourage independent laboratory experimentation. While such a phenomenon would have seemed to absolutely go against the law of energy conservation a number of years ago, we now recognize that the potential for extracting energy out of so called empty space is in fact a reality".

Not So Empty Space:

Puthoff, a PhD. from Stanford University, believes that a new, nonpolluting energy source may be achieved by tapping the force random fluctuations jostling atomic particles within a vacuum. Scientists now know that "empty" space seethes with what are called vacuum fluctuations: huge amounts of energy that suddenly burst forth, jiggling particles to and fro. Puthoff has developed his own theory, zero point energy, in an attempt to tap the abundant power found in the vacuum of space. He and associates in a new company, Jupiter Technologies, may soon try to manufacture zero-point energy machines.

DePalma described his N machine and outlined a theory to explain its workings in a paper, "On the Possibility of Extraction of Electrical Energy Directly From Space", published in the British science journal, Speculations in Science and Technology (Sept 1990, Vol 13 No 4). So far, the scientific establishment either has ignored DePalma's controversial claims or remains unaware of them.

Patent *Not* Pending:

No one has ever obtained a patent for an N machine in the U.S., although in the San Francisco area alone, there are some 200 patent applications relating to such devices. The U.S. Patent office automatically denies a patent to any gizmo which purports to produce more energy than it consumes, on the grounds that its personnel are not equipped to evaluate such claims. DePalma is quick to point out that the N machine is not a perpetual motion machine, that mythical contraption long sought by frustrated inventors. "The perpetual motion machine is only supposed to run itself. It could never put out five times more power than is put into it. Perpetual motion schemes used conventional energy sources, whereas the N machine is a new way of extracting energy from space".

Other scientists-inventors who attempted to build and operate free energy machines have been intimidated and harassed by the U.S. government. At least one inventor had his device confiscated by the Defense Department on the grounds that its free energy technology endangered national security interests. This inventor was put under a gag order, so that he could not even tell the press that his N machine had been confiscated. This is ironic when one considers that the idea for the N machine came directly from a famous experiment performed by Michael Faraday in 1831.

U.S. Not Interested:

The U.S. energy monopoly, which pushes for the development of oil, gas, coal, and nuclear power--while defunding solar energy and other nonpolluting alternatives--apparently does not want to see free energy emerge as a viable option.

Meanwhile, other countries, notably India and Japan, are vigorously pursuing what might prove to be a technological breakthrough. (is this yet one more example of the Invented in USA/Made in Japan" syndrome, the outcome of American shortsightedness and vested interests?) In India, eminent engineer Paramahamsa Tewari is currently testing his invention, called the Space Power Generator (SPG), which essentially replicates DePalma's N machine. With 5 kilowatt total input, the SPG is reportedly yielding 30 kilowatt electrical output (correspondence to B. DePalma 8/13/90).

Tewari, a senior engineer with India's Department of Atomic Energy-Nuclear Power Corporation, also directs the Kaiga Project, India's largest atomic power facility, in Karnataka. He freely acknowledges his dept to DePalma, who has shared his experimental results with Tewari for many years. According to Tewari, "The electrical power generated by the Space Power Generator is indeed commercially viable and should be brought to the notice of the general public." He has urged India's Atomic Energy Commission to create an independent research group to advance free energy technology. Tewari also credits John Wheeler, the prominent American physicist and discoverer of the black hole, for his steady encouragement. Wheeler, who had been searching for a mathematical theory that would predict free energy, applauded Tewari for his efforts to develop such a theory, and the two scientists corresponded for several years.

Japanese Interest:

The Japan Science Foundation, under Japanese government auspices, awarded grants to two universities and one company to produce models of the N machine and to investigate how it works. Kazama Giken Corporation is commercially supplying small N machines for research and educational purposes . Another Japanese company, Panasonic/National, is also pursuing this technology. Shiuji Inomata, Ph D president of the Japan Psychotronic Institute and senior scientist at the Electrotechnical Laboratory in Ibaraki, has been instrumental in sparking the interest of Japan's scientific community in the N machine.

"One day man will connect his apparatus to the very wheelwork of the universe... and the very forces that motivate the planets in their orbits and cause them to rotate will rotate his own machinery," predicted Nikola Tesla, the Croatian born American electrical genius whose discoveries and inventions rival those of Edison. Proponents of the N machine believe that it taps directly into a primordial energy source, meshing with the wheelwork of the cosmos.

A Wrong Turn:

"Electrical engineering took a wrong turn 160 years ago," according to Tewari, referring to English scientist Michael Faraday's pioneering work of the world's first dynamo. In 1831, Faraday performed a series of experiments which led to the modern electric induction generator, having two moving parts--a rotor and a stator. Faraday moved a wire near the pole of a magnet, producing an electrical potential across the ends of the wire. This induction principle is used in all the electrical generators we use today. And that's precisely what Tewari means by a "wrong turn."

In that same year, 1831, Faraday also performed a simple yet ingenious experiment with a rotating magnetized conductor. The resulting phenomenon (free energy?) has yet to be explained in terms of conventional scientific theory.

By cementing a copper disc on top of a cylinder magnet, and rotating the magnet and disc together, Faraday created an electrical potential. After pondering this phenomenon for many years, he concluded that when a magnet is rotated, its magnetic field remains stationary. Thus, he reasoned, the metal of the magnet moves through its own field, and the relative motion is translated into electrical potential.

Farady's experiments led him to the revolutionary conclusion that a magnetic field is a property of space itself, not something attached to the magnet, which merely serves to induce or evoke the field.

A Prototype:

Known for over 150 years, the Faraday homopolar generator, as his contraption is called, has been viewed by a handful of visionary inventors as a basis for evoking the free energy latent in space. They see is as the prototype for a generator capable of providing its own motive power with additional energy to spare. When the world embraced Faraday's two piece induction generator, whose drawbacks include mechanical friction and electrical losses, the enormous potential of the Faraday homopolar generator was abandoned, in the opinion of free-energy proponents.

Following in Faraday's footsteps, DePalma in 1978 speculated that free energy could be tapped from the matrix of space simply by magnetizing a gyroscope. "I reasoned that the metal of the magnetized gyroscope moving through its own magnetic field, when rotated would produce an electrical potential between the axle and the outer edge of the rotating magnetized flywheel," he explains.

This insight led to his N machine, essentially a one piece rotating magnetized flywheel, "Instead of having a rotor and a stator, as do conventional generators, the n machine only has a rotor.. Half of the flywheel is the north pole; the other half is the south pole. One electrical contact is put on the axle, another contact is placed on the outer edge of the gyroscope, and presto, electricity is taken directly out of the magnet itself.

Idea Put to the Test:

For 150 years after Faraday's controversial experiment, no one bothered to see whether or not a rotating magnet generator would have to the same amount of work as a conventional induction generator in order to produce and identical power output. Then, in 1978, the aforementioned Sunburst homopolar generator was built. Tests determined that is output power greatly exceeded the input needed to run the machine, that it was much more efficient that an induction generator. Opinions differ as to the exact mechanisms by which the N machine generates energy. In 1977 Tewari created a minor sensation when he put forth the theory that space is filled with a dynamic medium whose swirling motion is the source of all matter and energy. In his Space Vortex Theory, more fully developed in his 1984 book, Beyond Matter, the Indian engineer inventor postulated that a void lies at the heart of the electron-- a void whose high speed rotation within a vacuum could produce energy from space. Tewari's theory is based on the assumption that the electron has a definite structure, and is not just a homogeneous "droplet of charge".

According to Tewari, the movement of "voids" in the spinning magnetized cylinder of his Space Power Generator liberates free energy out of the space between the machine's axis and the magnet. He readily admits that this sounds incredible, by the yardstick of known laws of physics. Tewari says he never would have developed his theory had he been trained as a physicist rather than as an engineer, since his ideas differ so radically from conventional physics.

"Tewari's explanation is perfectly possible," comments DePalma. "He is attempting to conceptualize what's happening between the atoms and where the energy is liberated."

Concept of Magnetism:

"My own approach," continues DePalma, "is that space is all around us like the sea of water the fish swim in. The only way you know it's there is to distort it in some way, and the simplest way to distort space is with a magnet." DePalma maintains that his own conception of magnetism as a distortion of a pre-existent homogeneous field is "the first new thought on the fundamental nature of magnetism since Oersted."

Having taught at MIT as a lecturer in physics for 15 years, DePalma grew increasingly dissatisfied with mainstream physics' explanation of the way things work. His current view of the universe would strike many conventional scientists as heresy.

For example, modern science says that energy is a constant substance in the universe, and that the conversion of energy from one form to another will lead to the heat-death of the universe eons from now. In contrast, says DePalma, "My cosmos is an open-ended universe, one in which energy can be evoked from space itself. All energy come from space," he maintains, "and there are various processes which can release this energy, the simplest of which is lighting a match or rubbing two sticks together."

Suppose you light a candle. The heat in the flame derives from the release of latent heat stored in the wax, according to the textbooks. Nonsense says DePalma. "The law of energy conservation is pure assumption," he insists. In his theory, the heat of a lit candle comes from space, and this substrate is slowly consumed by the energy of space flowing through it.

When you drive a car, the heat latent in the gasoline is extracted through burning, which propels the pistons. Right? Wrong says DePalma. His understanding of the process is that the gasoline-air mixture, catalyzed by an electric spark, acts as a "molecular antenna" to release energy from space. Heat energy thus release cooks or burns the substance which is evoking it in the first place, producing exhaust.

Likewise, when a magnet is rotated in the N-machine, DePalma theorizes, the electrical current comes from the space through which the magnet is drawing its energy, not from the magnet's mechanical rotation.

DePalma's approaches to other basic phenomena are equally unorthodox. In the mid-1970's he performed the "Spinning Ball Experiment," which purportedly demonstrated that a rotating object will fall faster or go higher than an identical non rotating object with the same initial velocity. If true, these results fly in the face of all knows physics. The experimental procedure is simple: Take a steel ball bearing (the kind used in a pinball machine), set it rotating, drop it, and measure how fast it falls. Compare its time with that of an identical but non-rotating control ball.

DePalma explained his anomalous results in terms of free energy added to the motion of the rotating object. These and other experiments led him to formulate radical new concepts of rotation, gravity, inertia and motion in general building on the work of pioneers in the field. He published his findings on the Spinning Ball Experiment in the British Scientific Research Association Journal in 1976. He also outlined the Spinning Ball Experiment to Dr. Edward Purcell, a Harvard physics professor, one of the most eminent experimental physicists at that time. According to DePalma, Purcell, after contemplating the experiment for several minutes, blurted, "This will change everything."

Applying New Technology:

"Applied physics is not engraved in stone," says Don Kelly, president of the Space Energy Association (SEA/US), a group of engineers, scientists, and inventors dedicated to developing free-energy technology. Today's free-energy scene encompasses a bewildering array of devices, including N machine, Russian plasma generators, a Swill hybrid converter (combining free-energy components and solid state methods), permanent magnet motors (PMMs), the multiple-coil Hubbard Generator, and various hydrogen power systems.

A standout among the latter group is the "Enerex" H20 unit invented by Yoshiro Nakmatsu, known as the "Edison of Japan." This prolific inventor--

father of the floppy disk--claims that his pollution free Enerex unit runs on tap water alone and can generate three times as much power as a standard gasoline engine. An H20 splitter, the Enerex produces hydrogen as the working fuel.

Kelly notes that Germany, Switzerland, Japan, Korea and the Netherlands all have active free-energy research associations, with which SEA/US exchanges information. Nevertheless, he feels that the nascent free energy technology faces opposition in the United States from government agencies, academics, and vested industrial interests. Kely envisions free energy eventually gaining acceptance and application through a grassroots movement of do-it-yourselfers around the country. SEA/US publishes an interesting quarterly newsletter, available to members. (Space Energy Association/U.S. P.O. Box 11422, Clearwater, FL 34616; phone 813-441-3923; membership dues are $35 per year).

Economics of the N machine:

DePalma Energy Corporation has not sold a single machine yet. To build an N machine by hand, the company charges around $500,000. Bruce DePalma claims that, if mass produced, the cost of his machine would drop to a mere $400 to $500. He points out that a typical 100 kilowatt AC generator costs a little over $100,000, and add that an N machine putting out the same amount of power could be manufactured for one third to one half the cost in regular production. His goal is to set up technology sharing agreements with clients who would manufacture his machine.

Surveying the variety of electro-magnetic free energy units available or on the drawing boards, Don Kelly concluded that most of them are plagued by fuzzy applied physics, lack of technical and financial support, and "a distinct cost effectiveness problem." However, Kelly singled out the DePalma N machine as "the mainstay for this F/E category" and "the best of the F/E units for its potential today." He gave the N machine a high "KISS rating" (KISS=Keep it simple and stupid) due to the machines simple, one-piece rotor performs better than today's conventional two piece generators," says Kelly.

Reference Article - A Cheap, Inexhaustible, Self-Sustaining and Non Polluting Source of Energy, written by Richard Walters, People Magazine / Energy/New Ideas section: Subtitle: A promising new alternative energy source, neglected in the U.S. advances in the Far East. Extracted from the Internet January 8, 2019 @ https://www.bibliotecapleyades.net/ciencia/secret_projects/project014.htm[xi]

The Wireless Power Grid: More Than A 100 Years In The Making

If you can imagine a massive, horrifying beast with some 6 million miles of tentacles that costs up to $33 billion to feed and $5 trillion to replace, then you can imagine the U.S. electricity grid.

But it's a beast that can possibly be vanquished, finally.

How? Magnetic induction and resonant coupling.

Imagine a future where you can charge your electric vehicles while in motion using a charging mechanism built into the road on which it drives. Where the electrical grid is no longer reliant on power lines, utility poles, or expensive transformers and underground cables. A future where power companies stop chopping down trees that threaten nearby power lines.

This might sound like a sci-fi movie, but the future you just imagined is nearly here, in the form of magnetic inductive coupling and resonant coupling.

Magnetic induction, or magnetic inductive coupling, courtesy of Nikola Tesla, is already a mainstay in small-scale technology such as wireless cellphone charging and wireless speakers. Transformers also use this technology, which allows energy to be transferred from one coil to another, but the coils can be only centimeters apart—any further and it won't work.

Resonant coupling works similarly to magnetic induction, but allows for a greater distance between the two coils.

Naysayers abound, but magnetic induction and resonant coupling is not just possible on a large scale—it's inevitable. And none too soon, as the requirements for distributing electric power continue to change with the gaining presence of renewable energy sources and increasing fears of grid security and unreliability.

It may seem like a mighty big leap to go from wireless cellphone charging to a wireless power grid, but mankind has been known to make some huge jumps, courtesy of unconventional blue-sky thinkers, MIT grads, and deep-pocketed corporations.

To Boldly Go Where No Man Has Gone Before

It shouldn't come as a surprise to learn that some of the most creative tech ideas have come from our creative visionaries in Hollywood. They've generated some spectacular ideas that have forever changed the world.

Hollywood has already gone where no man has gone before, and the rest of the world has followed.

Fifty-one years ago, when "Star Trek" burst onto the television scene, minds were blown with the then-silly notion of a handheld computer. At the time, it seemed rather farfetched; the personal computer hadn't been invented yet, let alone laptops and tablets. But here we are, with iPods and tablets and Chromebooks and smart watches. The sci-fi wonder was several iterations ahead of reality, and now the rest of the world has arrived, albeit lagging by a few decades.

Star Trek also thought up the universal translator, which was probably introduced solely to explain why everyone on the show spoke English even though they were supposed to be from different planets. Regardless of the motive behind this invention, that universal translator—then only a finished product idea that lacked even a hint of how to make it happen—is finally here. We have had Google Translate (granted, it's not perfect) for quite some time now, but just last month Google announced its Pixel Buds, which can translate 40 languages in real time.

Hollywood's thought contributions don't end there. Next on the list is the holodeck. While it first appeared in "Star Trek: The Next Generation", it was actually thought up in the original series along with all the other cool gizmos, but was never shown on screen. If you're under the age of 35 and have no idea what a holodeck is, just ask Siri, Bixby, or Alexa (yeah, the talking computer was a Star Trek invention, too) and she'll explain, perhaps by showing you today's version of the holodeck: virtual reality goggles.

Other Star Trek high-tech gadgetry also makes the list of up-and-coming reality, and include the communicator badge, which is currently in prototype phase, and the tractor beam, which is in development. And don't tell your kids just yet, but even the hypospray is here.

The most vital industry information will soon be right at your fingertips

And it's not just Star Trek. Twenty years its junior, the "Back to the Future" franchise conceptualized even more things that were nonexistent then, but real things now. It thought up drones, hands-free gaming, mobile payment technology, hoverboards, biometric devices, wearable technology, video calling, and probably the most relevant for our readers, the DeLorean—a car that can run on garbage and duals as a time machine.

Well, we might not have a car that can run on actual garbage, but eco-friendly hydrogen-powered cars are indeed here. As for the rest of the tech mentioned above.... yeah, we've got that. As for the time machine, that's probably a bit further out.

The present is theirs; the future, for which I really worked, is mine. — Nikola Tesla

It's not just Hollywood scriptwriters who serve as the artistic muse for MIT's high-tech hotshots. Wireless electricity was the brainchild of Nikola Tesla, who as early as the 1890s had grand ideas of a global wireless power grid—and some university students now think they're well on their way to making his vision a reality.

A decade ago, MIT researchers proved that it was indeed possible to wirelessly power a light bulb more than two meters away-, and while it's not exactly the finished product that Nikola Tesla had in mind, he would have been proud; it was the first such feat and a milestone for wireless electricity.

More progress continues to be made. This year, senior study author and professor of electrical engineering Shanhui Fan said in a June interview for Stanford News that they have developed a way to wirelessly charge moving objects—a clear precursor to wirelessly charging not just stationary objects but electric vehicles while in motion. The technology could make alleviate the concerns that currently exist in the automobile marketplace about how far an electric car can travel without a charge—a major hurdle to making EVs more prevalent on the road.

"We still need to significantly increase the amount of electricity being transferred to charge electric cars, but we may not need to push the distance too much more," Fan said, adding that his hope is that "you'll be able to charge your electric car while you're driving down the highway. A coil in the bottom of the vehicle could receive electricity from a series of coils connected to an electric current embedded in the road."

And this induction system, the "electric road of the future" if you will, is already being tested in France. Carmakers have yet to incorporate into their vehicles the necessary technology that would allow them to top up by pads under the road surfaces, but this technology is already being tried out in a test track near Paris. Full-scale introduction of this tech is reportedly ten years out.

"Maybe 10 years is a good timescale for this technology," Virginie Maillard, a senior Renault EV engineer said in an interview with Autocar Magazine. "We have to design cars and the road network to accept it."

Back in 2012, Pike Research acknowledged the progress that had been made, but stated that the technology to replace existing utility poles with wireless power was decades away. But today's growing fears about grid security and grid reliability may speed along its progression.

United States President Donald Trump designated November 2017 as Critical Infrastructure Security and Resilience Month. The Department of Energy approved two transmission projects this year, with the last one gaining approval yesterday: The $1.6 billion Northern Pass Transmission Line project will bring hydropower from America's northern neighbor, Quebec, by tying the American grid to Canada's and lowering the carbon footprint in the region.

The regulatory approval of two major transmission projects is a clear sign that the new administration is working to improve the nation's infrastructure and streamline the federal permitting process, which may mean less regulatory hurdles for grid projects going forward—projects like a wireless grid, for instance.

A truly wireless grid isn't something we're likely to see tomorrow, but it's likely to be a reality someday in the fairly near future, and the power industry from power companies to transformer and power line manufacturers—and everything in between-—should at least consider that future.

If Star Trek and other sci-fi ideas are any indicator, we may very well see the beast defeated in our lifetime.

Reference[xii] Article – The Wireless Power Grid: More Than A 100 Years In The Making. Written and posted By Julianne Geiger - Nov 19, 2017, 5:00 PM CST. Geiger is s a veteran editor, writer and researcher for US-based Divergente LLC consulting firm, and a member of the Creative Professionals Networking Group. Extracted from the Internet @ https://oilprice.com/Energy/Energy-General/The-Wireless-Power-Grid-More-Than-A-100-Years-In-The-Making.html

New Energy Sources and Inventions

In the late 1880's, trade journals in the electrical sciences were predicting "free electricity" in the near future. Incredible discoveries about the nature of electricity were becoming commonplace. Within 20 years, there would be automobiles, airplanes, movies, recorded music, telephones, radio, and practical cameras.

For the first time in history, common people were encouraged to envision a utopian future filled with abundant modern transportation and communication, as well as jobs, housing and food for everyone. So what happened? Where did the new energy breakthroughs go? Was this excitement about free electricity all just wishful thinking that science eventually disproved?

Current State of Technology

The answer is no. Spectacular new energy technologies were developed right along with all the other breakthroughs. Since then, multiple methods for producing vast amounts of energy at extremely low cost have been developed. None of these technologies have made it to the open consumer market, however. Why this is true will be discussed shortly. First, here is a short list of new energy technologies. The common feature connecting all of these discoveries is that they use a small amount of one form of energy to control or release a large amount of a different kind of energy.

Radiant Energy

Nikola Tesla's Magnifying Transmitter, T. Henry Moray's Radiant Energy Device, Edwin Gray's EMA Motor, and Paul Baumann's Testatika Machine all run on Radiant Energy. This natural energy can perform the same wonders as ordinary electricity at less than 1% of the cost. It does not behave exactly like electricity, however, which has contributed to the scientific community's misunderstanding of it. The Methernitha Community in Switzerland currently has 5 or 6 working models of fuelless, self-running devices that tap this energy. [More]

Permanent Magnets

Dr. Tom Bearden has two working models of a permanent magnet powered electrical transformer. It uses a 6-watt electrical input to control the path of a magnetic field coming out of a magnet. By channeling the magnetic field, first to one output coil then a second repeatedly and rapidly, the device can produce a 96-watt electrical output with no moving parts. Multiple inventors have working mechanisms that produce torque from permanent magnets alone. [More]

Super-Efficient Electrolysis

Water can be broken into hydrogen and oxygen using electricity. When water is hit with its own molecular resonant frequency, it collapses into hydrogen and oxygen gas with little electrical input. Hydrogen fuel can drive engines (like in your car) for the cost of water. [More]

Cold Fusion

Though initial claims were debunked, cold fusion is very real. Not only has excess heat production been repeatedly documented, but also low energy atomic element transmutation has been catalogued, involving dozens of different reactions. [More]

There are dozens of other systems

Many are viable and well tested. But this short list is sufficient to make the point: new energy technology is here. It offers the world pollution-free energy abundance for everyone, everywhere. It is now possible to stop the production of "greenhouse gases" and shut down the nuclear power plants. Transportation and production costs for just about everything can drop dramatically. Yet all these wonderful benefits that can make life on this planet so much easier and better for everyone have been postponed for decades. Why? Whose purposes are served by this postponement?

Four Invisible Forces. There are four forces that have worked together to create this situation.

The wealthiest families and their central banking institutions are the first force. Their motivations are greed and the need to control almost everything except themselves. Their plan is to eventually control all of the resources of the world, and thereby control everyone's life through the availability of all goods and services. An independent source of wealth (new energy device) in the hands of every person in the world ruins their plans for world domination. They don't want any competition. The weapons they have used to enforce this include intimidation, "expert" debunkers, buying and shelving of technology, and murder of inventors. They have also promoted the scientific theory that states free energy is impossible (laws of thermodynamics).

The second force is national governments. The problem here is related to the maintenance of national security. There is a constant jockeying for position and influence in world affairs, and the strongest party wins. In economics, it's the golden rule: "The one with the gold makes the rules." Unlimited energy on this planet will lead to an inevitable reshuffling of the balance of power. Everybody will want it, and at the same time, want to prevent everyone else from getting it. So, national governments' motivations are self-preservation. Their weapons include preventing the issuance of patents based on national security grounds, harassment of inventors with criminal charges, tax audits, threats, phone taps, arson, theft, and a host of other intimidations which make the business of building and marketing a new energy machine practically impossible.

The third force consists of deluded inventors and con men. On the periphery of the extraordinary scientific breakthroughs that constitute real new energy technologies, lies a shadow world of unexplained anomalies, marginal inventions and unscrupulous promoters. The first two forces have constantly used the media to promote the worst examples of this group, to distract the public's attention, and to discredit real breakthroughs by associating them with the frauds. So, the third force is delusion and dishonesty. The motivations are self-aggrandizement, greed, want of power over others, and a false sense of self-importance. The weapons used are lying, cheating, self-delusion and arrogance combined with bad science.

The fourth force operating to postpone the public availability of new energy technology is all of the rest of us. It may be easy to see how narrow and selfish the motivations of the other forces are, but actually, these motivations are still very much alive in each of us as well. Like the wealthiest families, don't we each secretly harbor illusions of false superiority and want to control others instead of ourselves? Also, wouldn't you sell out if the price were high enough? Or like the governments, don't we each want to ensure our own survival? Or like the deluded inventor, don't we trade a comfortable illusion once in a while for an uncomfortable fact? Or don't we still fear the unknown, even if it promises a great reward?

All four forces are just different aspects of the same process. There is really only one force preventing the availability of new energy technology, and that is unspiritually motivated behavior. New energy technology is an outward manifestation of divine abundance. It is the engine of the economy of an enlightened society, where people voluntarily behave in a respectful and civil manner toward each other. Unspiritualized humans cannot be trusted with new energy. They will only do what they have always done, which is to take merciless advantage of each other, or kill each other and themselves in the process.

The Opportunity. What is new is that you and I can communicate with each other now better than at any time in the past. The Internet offers us, the fourth force, an opportunity to overcome the combined efforts of the other forces preventing new energy technology from spreading. What is starting to happen is that inventors are publishing their work, instead of patenting it and keeping it secret. More and more, people are "giving away" information on these technologies in books, videos and websites. While there is still a great deal of useless information about new energy on the Internet, the availability of good information is rising rapidly.

All of us constitute the fourth force. If we stand up and refuse to remain ignorant and action-less, we can change the course of history. Only mass action can create the world we want. The other three forces will not help us put a fuelless power plant in our homes. New energy technology will change everything about the way we live, work and relate to each other. It obsoletes greed and fear for survival. But like all exercises of spiritual faith, we must first manifest generosity and trust in our own lives.

New energy technology is here. It has been here for decades. Communications technology and the Internet have torn the veil of secrecy off of this remarkable fact. People all over the world are starting to build new energy devices. The bankers and governments don't want this to happen, but can't stop it. Tremendous economic instabilities and wars will be used to distract people's attention from the new energy movement. There will be essentially no major media coverage of this aspect of what is going on.

Western society is in many ways spiraling toward self-destruction due to the accumulated effects of greed and corruption. New energy technologies cannot stop this trend. If, however, you have a new energy device, you may be better positioned to support the transition that is underway. The question is, who will ultimately control the emerging world government, the first force, or the fourth force? Those who choose the fourth force may live to see the dawn of the world of new energy. I challenge you to be among the ones who do so.

Reference[xiii] Article - New Energy Sources and Inventions. This is an extract of the book titled *New Energy Sources* that was written by Peter Lindemann, D. Sc. and summarized and edited by Fred Burks https://www.wanttoknow.info/newenergysources

Cheap power: An overnight revolution

In a couple of weeks the world could be forever changed by power too cheap to meter. Or not.

Every now and then along comes a technology that is revolutionary and changes everything. But a very few of these new technologies cause fast change. Mostly they seep out of the lab, into the arms of early adopters, and then ooze out into the world in general.

For example, the first internal combustion engine was (arguably) the Pyréolophore invented by Nicéphore and Claude Niépce in 1807, but car ownership wasn't really commonplace until the first decade of the 1900s.

So, while the internal combustion engine is one of the key technological achievements that define Western culture, it took more than a century for the impact of the technology to become widespread. As technological revolutions go, that one was pretty slow.

The "Singularity is Near" website has a number of charts illustrating the pace of technology adoption. One of the most interesting concerns the mass use of inventions, which shows the number of years required for various technologies to be adopted by one-quarter of the U.S. population.

We can see from this that electricity transmission and use, which was developed in 1873, took 46 years to reach a mass market, while the telephone (1876) took only 35 years. The radio (1897) took only 31 years, television (1926) took 26, the PC (1975) just 16 years, the mobile phone (1983) a mere 13 years, and the Web (1991) a blisteringly fast seven years.

Now, in IT we're used to change and revolutionary (or rather what appear to be revolutionary) changes at that. I won't labor the statistics of technology evolution other than to point out how the cost of rotating disk storage has fallen from 158 bits bytes per dollar in 1956 to 25GB per dollar today!

All the forgoing was just to provide perspective and frame what might be about to happen: a technological advance that in terms of societal impact could be greater and far faster than pretty much anything else we've witnessed in human history. That thing could come ... and I'm not kidding ... in the shape of a steam engine.

An Italian inventor named Andrea Rossi and his scientific consultant, physicist and emeritus professor, Sergio Focardi, have demonstrated a device called the E-Cat or Energy Catalyzer which, according to a 2008 patent application, involves "a method and apparatus for carrying out nickel and hydrogen exothermal reactions," with the production of copper as a result.

The device is said to work by heating hydrogen to an "ignition temperature" using an external heat source, after which a catalyst, which has yet to be explained, causes the hydrogen atoms to "penetrate" the nickel and transform it into copper, producing energy in the process -- essentially a nuclear fusion reaction -- that is self-sustaining (i.e. the external heat source can be removed and the device will continue to function).

Water fed into the reaction chamber comes out as steam with which you could drive a turbine, and voila! You have a generator. Or you could use it for motive power. You could also use the heat to drive a Stirling Engine, but for whatever reason, this option hasn't been much discussed.

The whole idea of generating power from nuclear fusion has been, if you'll excuse the pun, a "hot" topic for a long time. Most attempts to build fusion power generators have been mainly "Big Science" experiments costing millions of dollars, such as the National Spherical Torus Experiment, the International Thermonuclear Experimental Reactor, the Tokamak Fusion Test Reactor and the Polywell.

All of these designs rely on the creation of extreme environments where a high-temperature plasma (a very hot "gas" of ionized particles) is confined by a powerful magnetic or electrostatic field. This is not engineering you do casually. Or cheaply.

But hot fusion is not what the E-Cat does and while much of the commentary on this device characterizes it as "cold fusion," Rossi claims that it isn't actually cold fusion at all but involves a Low-Energy Nuclear Reaction (I can't figure out what the difference between cold fusion and LENR might be from the research I've done).

Cold fusion, a nuclear fusion reaction at low energies or temperatures, is the alternative to hot fusion and includes Muon-catalyzed fusion and "generally cold, locally hot" fusion.

Experiments using these methods of cold fusion have, in common with hot fusion, so far not shown a net energy gain, although the experiments have been much cheaper.

A much-celebrated claim for a working cold fusion reaction became a huge brouhaha in 1989 when two established scientists, Martin Fleischmann, one of the world's leading electrochemists, and Stanley Pons, also a respected electrochemist, claimed to have observed cold fusion in a table-top experiment.

Alas, after many attempts by other researchers to duplicate the findings, as well further work by Fleischmann and Pons at a cost of more than $44 million, it transpired that their results could not be replicated and they and their experiment were discredited ... as was the idea of cold fusion.

But all of these failures in both hot and cold fusion haven't stopped research because, if nuclear fusion can be made to work, we could have essentially unlimited energy for next to no cost beyond infrastructure costs (generator production, transmission, facilities, manpower, waste management and so on). So far, no joy. Not one experiment has demonstrated what is termed "over unity" output -- that is, the ability to produce more power than is input.

It is in this netherworld of physics that Rossi's system has appeared and, no surprise, along with it, a three-ring circus of media, science and speculation. At the center of the show is the E-Cat device which has been demonstrated a number of times and appears to be very simple.

One of the demonstration attendees, Bologna physics professor Giuseppe Levi, described as "an expert on nuclear physics, energy physics and sub-nuclear physics," has publicly defended Rossi and Focardi as well as the project.

There's now quite a furor over how the E-Cat might really work (or not), with some critics predicting that eventually the system will not perform as claimed and accusing the inventors of being deluded, or worse, frauds.

The results of the demonstrations have received a tremendous amount of analysis and commentary but, for now, the metaphorical jury is out.

Rossi has another demonstration scheduled in Bologna, Italy, for Oct. 28 for which he has partnered with a U.S. company, AmpEnergo, to build what is claimed to be a 1MW plant in a shipping container.

So, here's the question: Let's assume Rossi's E-Cat works. What then?

From the demonstrated prototypes it appears that you could build E-Cats small enough to power a car or a house. Bundle a lot of them together and you could power a truck, a ship or an office block. Imagine a data center where each row of racks has its own really cheap power generator.

Now you have a world where oil only matters as a raw product for things like plastics so the oil economy as we know it could be dismantled within a few months. Production costs for anything would fall. The power grid would become obsolete. Power stations of all kinds would no longer be an environmental problem. The balance of economic power worldwide would change and for example, OPEC would become a historical footnote.

The only risk, assuming that the E-Cat doesn't become horribly radioactive after extended operation or produce some other kind of hazardous byproduct, could be global thermal pollution from so many power generators (if they are very cheap and not dangerous then niceties such as minimizing waste heat would be ignored).

We could see a world where ubiquitous power generation is so cheap it wouldn't be worth metering (as a consequence, Rossi would become the wealthiest man in the world, assuming that all of the vested interests in the existing oil and power economies didn't have him bumped off).

It remains to be seen whether this is really all some kind of mistake, which seems unlikely, or a hoax, which seems equally implausible because, if it is all bogus, then there's no obvious upside for Rossi or any of the others involved.

So, Oct. 28 will be a big day. If the demonstration goes ahead as planned either we're going to be really disappointed or we'll be on the brink of something that will change the world forever.

You thought the adoption of the Web was fast? This could change everything overnight.

Reference Article - Cheap power: An overnight revolution. Written by Mark Gibbs whom is an author, journalist, and contributing editor to NetworkWorld from IDG. This article posted October 2011 at https://www.forbes.com/sites/markgibbs/2013/05/20/finally-independent-testing-of-rossis-e-cat-cold-fusion-device-maybe-the-world-will-change-after-all/#69b93c3924dd [xiv]

The Coming Era of Unlimited, Free and Clean Energy

In the 1980s, leading consultants were skeptical about cellular phones. McKinsey & Company noted that the handsets were heavy, batteries didn't last long, coverage was patchy, and the cost per minute was exorbitant. It predicted that in 20 years the total market size would be about 900,000 units, and advised AT&T to pull out. McKinsey was wrong, of course. There were more than 100 million cellular phones in use in 2000; there are billions now. Costs have fallen so far that even the poor — all over world — can afford a cellular phone.

The experts are saying the same about solar energy now. They note that after decades of development, solar power hardly supplies 1 percent of the world's energy needs. They say that solar is inefficient, too expensive to install, and unreliable, and will fail without government subsidies. They too are wrong. Solar will be as ubiquitous as cellular phones are.

Futurist Ray Kurzweil notes that solar power has been doubling every two years for the past 30 years — as costs have been dropping. He says solar energy is only six doublings — or less than 14 years — away from meeting 100 percent of today's energy needs. Energy usage will keep increasing, so this is a moving target. But, by Kurzweil's estimates, inexpensive renewable sources will provide more energy than the world needs in less than 20 years. Even then, we will be using only one part in 10,000 of the sunlight that falls on the Earth.

In places such as Germany, Spain, Portugal, Australia, and the Southwest United States, residential-scale solar production has already reached "grid parity" with average residential electricity prices. In other words, it costs no more in the long term to install solar panels than to buy electricity from utility companies. The prices of solar panels have fallen 75 percent in the past five years alone and will fall much further as the technologies to create them improve and scale of production increases. By 2020, solar energy will be price-competitive with energy generated from fossil fuels on an unsubsidized basis in most parts of the world. Within the next decade, it will cost a fraction of what fossil-fuel-based alternatives do.

It isn't just solar production that is advancing at a rapid rate; there are also technologies to harness the power of wind, biomass, thermal, tidal, and waste-breakdown energy, and research projects all over the world are working on improving their efficiency and effectiveness. Wind power, for example, has also come down sharply in price and is now competitive with the cost of new coal-burning power plants in the United States. It will, without doubt, give solar energy a run for its money. There will be breakthroughs in many different technologies, and these will accelerate overall progress.

Despite the skepticism of experts and criticism by naysayers, there is little doubt that we are heading into an era of unlimited and almost free clean energy. This has profound implications.

Why the world still uses coal

Coal is dirty - so why are we still using it for energy? (Jorge Ribas and Julio Negron/The Washington Post)

First, there will be disruption of the entire fossil-fuel industry, starting with utility companies — which will face declining demand and then bankruptcy. Several of them see the writing on the wall. The smart ones are embracing solar and wind power. Others are lobbying to stop the progress of solar power — at all costs. Witness how groups in Oklahoma persuaded lawmakers to approve a surcharge on solar installations; the limited victory that groups backed by the Koch brothers won in Arizona to impose a $5 per month surcharge; and the battles being waged in other states. They are fighting a losing battle, however, because the advances aren't confined to the United States. Countries such as Germany, China, and Japan are leading the charge in the adoption of clean energies. Solar installations still depend on other power sources to supply energy when the sun isn't shining, but battery-storage technologies will improve so much over the next two decades that homes won't be dependent on the utility companies. We will go from debating incentives for installing clean energies to debating subsidies for utility companies to keep their operations going.

The environment will surely benefit from the elimination of fossil fuels, which will also boost most sectors of the economy. Electric cars will become cheaper to operate than fossil-fuel-burning ones, for example. We will be able to create unlimited clean water — by boiling ocean water and condensing it. With inexpensive energy, our farmers can also grow hydroponic fruits and vegetables in vertical farms located near consumers. Imagine skyscrapers located in cities that grow food in glass buildings without the need for pesticides, and that recycle nutrients and materials to ensure there is no ecological impact. We will have the energy needed to 3D-print our everyday goods and to heat our homes.

We are surely heading into the era of abundance that Peter Diamandis has written about — the era when the basic needs of humanity are met through advancing technologies. The challenge for mankind will be to share this abundance, ensuring that these technologies make the world a better place.

Reference[xv] Article - The coming era of unlimited, free and clean energy. Written by Vivek Wadhwa whom is a Distinguished Fellow and professor at Carnegie Mellon University Engineering at Silicon Valley and a director of research at Center for Entrepreneurship and Research Commercialization at Duke. His past appointments include Stanford Law School, the University of California, Berkeley, Harvard Law School, and Emory University. Extracted from the Internet January 13, 2019 @ https://www.washingtonpost.com/news/innovations/wp/2014/09/19/the-coming-era-of-unlimited-and-free-clean-energy/?utm_term=.3b25b38d1d7d

Wireless Power Is Coming

Companies like Energous and Ossia are developing 'true' wireless charging that works over the air so you can power up at a distance. But don't throw away your charging cable just yet.

Mobile World Congress is one of those events that make you realize the thin line between civilization and barbarism is the width of a single charging cable.

Anyone who's attended MWC (or any massive trade show) is familiar with the creeping dread caused by watching your phone's battery power seep away over the course of a very long workday. Tweet by tweet, photo by photo, text by text, the percentage points tick away, until you're scrambling in your bag for a battery pack amid a tangle of chargers and cables or searching the packed halls and press lounges for a free plug point—an oasis in the desert.

Imagine you never had to seek out an outlet, carry a backup battery, or stuff an array of cables into your bag again; that all your devices were being charged continuously, no matter where you went. Someday you may not have to imagine it—because companies like Energous and Ossia are working to make it a reality.

Nikola Tesla first envisioned a wireless power transmission system in the early 1890s; he even attempted to build an experimental station, Wardenclyffe Tower, which he hoped would be the first in a worldwide wireless system. Today, phones, smartwatches, and other devices sometimes come with inductive charging coils, which let them charge wirelessly through resonant inductive coupling. But this method relies on direct contact between the device and a charging pad—hardly the untethered future Tesla imagined.

That's where Energous and Ossia come in. Both focus on developing "true" wireless charging that works over the air (OTA) so you can power up at a distance. This is the wireless charging we were promised over a century ago. Could our dependence on cables or tethered charging solutions finally be coming to an end?

Why Today's Wireless Charging Disappoints

Though wireless charging has been around for years, it had a big boost last year, when Apple announced that its new phones had the feature. Qi, used by most phones (including iPhones), and the lesser-known PMA are the two forms of wireless charging used by current consumer-oriented devices. Both standards require one or more induction coils in a charging pad to transmit power and another in the device that's being charged.

There are some minor differences between the two standards, but they share the same limitations: the need for very close contact between the device and the charging pad, as well as proper alignment of the induction coils, for power to be transmitted.

The range can be slightly extended: The FCC Office of Engineering and Technology Laboratory Division recently approved an increase in distance of up to 20cm, potentially by using multiple charging coils. So far, though, no products currently utilize longer-range wireless charging for inductive charging coils.

Other potential problems with this type of wireless charging include heat buildup. "The circuitry in the wireless charger must be thoughtfully designed to prevent excess heat from building up in the charger and transferring to your smartphone," said Jack Norton, director of electrical engineering at Belkin. "Excess heat can lead to lower life spans for the wireless charger [and] your phone's battery."

The primary reason that wireless charging is slower than using a fast charge–enabled power brick is that a device will reduce its charging speed to manage heat and prevent damage to sensitive components. "Managing heat is definitely one of the challenges in wirelessly transferring power beyond 10 watts," said Charlie Quong, VP of product development and engineering for Zagg (Mophie).

Thermal dissipation can be managed in a variety of ways, including: passive airflow, which means putting holes or vents in strategic locations to transfer heat away from the device being charged; cooling fans; heat sinks made of metal or thermal glue; larger housing; specialized material; and more efficient component layouts. But even when you mitigate heat, inductive-charging speed for phones is limited. According to Pablo Valentin, global product manager at Belkin, "The fastest output realized in the market right now is 15W, both from charger side and phone side."

That means a device such as the Samsung Galaxy S9+$688.00 at Amazon, which is fast charge-compatible, can juice up to 50 percent in 1 hour and 34 minutes. That's not quite as fast as regular fast charging using an adapter (35 minutes to 50 percent), but it's a big step toward making wireless charging fast and convenient.

For larger devices such as laptops, which have more surface area and ventilation options, wireless charging can go to 30W or more. But when it comes to phones, you're still likely looking at a charging speed of 10W to 15W for most wireless charging pads, with most of them dropping down to slower wattages (5W) to prevent heat buildup. That's quite a bit slower than a 27W Qualcomm Quick Charge 4.0+ power adapter or 29W USB-C Power Adapter for the iPhone X$999.00 at Verizon Wireless, for example.

Another and perhaps more significant downside of inductive wireless charging involves the size of the charging coils. "There are different coil designs that can be implemented, depending on the application, but the size of the coil must be optimized and calibrated correctly for a wireless charging system to work properly," said Quong.

Charging coils can be made to fit on small devices such as Bluetooth earbuds, but such coils are uncommon. According to Gary Gao, wireless charging leader for Anker, the most common coil design is 50mm in diameter as required by the Qi standard.

Finally, the biggest limitation of Qi wireless charging is the paucity of the kinds of devices on which it can be implemented. "Right now, if you look at wireless charging, you see it on your phone primarily, and a few other random devices like the smartwatch from Apple," Energous President and CEO Stephen Rizzone said. "But it doesn't really get beyond that, and part of that is because of the limitations of that technology."

What Does This Look Like?

Though other companies are exploring this space, the Energous and Ossia wireless charging standards come closest to matching Nikola Tesla's vision. The two technologies operate using different radio frequencies, but they both let you charge a device from a distance, transmitting power over the air without using cables or inductive charging coils.

Energous: The Energous wireless charging technology works with the company's proprietary WattUp transmitter and receiver. It operates similarly to Wi-Fi by transmitting power in the 900MHz frequency to the receiver, which converts the RF signal into DC power.

Energous uses beamforming for the mid-field and far-field transmitters: a directional antenna array sends energy directly to the receiver. In the case of obstacles, it can reflect energy beams off walls and ceilings, although this results in some efficiency loss. The technology breaks down into three different types of wireless charging: near field, mid field, and far field, with varying ranges.

WattUp near-field charging requires a transmitter that can plug or be embedded into existing electronics—laptops, game consoles, tablets, and other devices. It's certified under Part 18 of the FCC requirements: That means it can output up to 10W of power because of the closer proximity to the transmitter. Rizzone anticipates WattUp charging higher-drain devices, such as phones, while still being able to charge any other device with the same charging circuit.

The mid-field transmitter is certified under Part 18 of the FCC's rules, so it can send large power outputs at a distance of 2 to 3 feet, with the standard use case being a desktop setup. Energous envisions it being designed into a desktop speaker and charging multiple devices placed around it. Far-field is the longest range currently possible: 15 feet. A typical use case would be a living room.

In each implementation of near-field, mid-field, and far-field charging, Rizzone imagines close integration with existing consumer products. "You're not going to want to have a big router-type-looking thing," he said. "Instead, you would primarily put that technology into the bezel of a television, [or] it could be a sound bar, it could be a few different things that are already in the room."

In addition to charging at a distance, each transmitter would be supported by a software layer (app- or web portal-based) to allow power management of devices connected to the various WattUp transmitters. You'll be able to control which devices are charged and to prevent your transmitter from charging your neighbors' devices. The mid-field and far-field transmitters also allow you to set device priorities and rules for charging and authentication and to configure charging zones. The short-range transmitter is more traditional: You can't set priorities or charging rules, but once you drop the device on the charging pad, you're able to view its battery level and expected time to charge fully.

The transmitter "sees all the different Bluetooth devices that are within the room," said Rizzone. "It identifies which ones are close enough to be charged and also identifies [whether] they're authorized on the network, and then it actually monitors, with sensors, what the user's typically doing."

This type of software-based power management is vital, Rizzone said, because "under Part 15 rules, you can only send one watt of power out. That means that at a foot away, you're going to get ... one milliwatt or lower. So a very, very small amount." (FCC Part 18 doesn't limit the amount of power).

With the widespread adoption of WattUp transmitters, though (however far off that might be), you'd be able to go to a coffee shop and even travel without having to carry adapters or cables: Various transmitters in public and private spaces would top up your various devices as you need them and according to your usage patterns.

Ossia: Energous is focused on consumer-level wireless charging, but Ossia has a bigger scale in mind.

As with Energous, the company's technology is similar to Wi-Fi and uses the 2.4GHz spectrum (it's considering the 5.8GHz spectrum for the future) under its proprietary "Cota" standard. Ossia's receiver sends a low-power beacon signal to the Cota Power Transmitter. The transmitter then directs power along the same pathway, with the receiver converting the RF signal to DC power. That allows Ossia's transmitter to send power to multiple Cota-enabled devices simultaneously and to send power to objects in motion. It also effectively bypasses objects that are in the way, since the signal doesn't pass through people, animals, or other organic matter.

The current implementation of Ossia's technology is currently showcased by two prototype devices—the Cota Forever Battery and the Cota Tile. The Forever Battery is an AA or AAA battery that wirelessly receives power from a Cota Power Transmitter on the 2.4GHz band. The Cota Tile is a transmitter capable of delivering several watts of power to Cota-enabled devices close to it and milliwatts at a distance of 30 feet. It operates at a greater range than Energous's far-field transmitter, but this requires the transmitter to be larger.

Wireless power ossia cota "We know that at a distance, signals drop off very quickly, and we don't really get much power in a far field," said Hatem Zeine, Ossia founder CTO. He explained that the company designs antennas and transmitters to be large enough to extend range of the near field.

Zeine was keen to point out that for consumers, the key to providing useful wireless power isn't necessarily to transmit a high degree of power to one or two devices. "The idea is that power has to be continuous, no matter how fast you go," he said. "This system [the Cota Tile] is running 2.4GHz with 20 watts coming out. At about a meter, we're receiving over 6 watts of power, and at two meters we're talking about a few watts, 2 to 3 watts, and then at the end of the room we're talking about 1 watt. So now, suddenly ... every cubic inch of this room has power."

In other words, while a Cota Tile may take a long time to power a phone except at close range, it's more than capable of keeping every device in a living room topped up—from game controllers to security cameras.

The method for connecting devices to the Cota Tile is similar to traditional Bluetooth or Wi-Fi pairing. The setup allows you to control which devices get charged and manage them as they charge. "If it's an intelligent device like this [Zeine indicates a phone], then you can basically go into the system here and say, 'Add me to the power network,' and it will talk to [the Cota Tile]," said Zeine.

The Cota Cloud software is able to define the priority of devices that get power and show the basic power usage of the device. When using the Cota Tile on a large scale, the administrator can set up and manage users. With prioritized charging, the Cota Tile first charges devices with lower battery levels; the Cota receiver communicates with the transmitter about how much power it needs. When there are no devices within range that need to be charged, the Cota Tile goes into sleep mode, so it's not wasting energy or transmitting power to devices that are already topped up.

The Cota Tile is able to power a variety of devices in a room because it doesn't have to operate on line of sight. A pair of Cota Tiles are twice as efficient and emit four times the power of a singla Cota Tile. So in a warehouse, for example, numerous Cota Tiles can join together to power hundreds or thousands of compatible devices.

"It's unique, because when you put two or maybe even four together, and you can envision it in the ceiling, that amount of power can power a coffee shop," said Zeine. "It's a legitimate deployment commercially, where we could power multiple smartphones at the same time. ... The other unique thing about having multiple tiles is they don't have to be right next to each other. You can space them apart and therefore create a [continuous] power situation." Like, say, walking along a corridor in an airport.

Since Ossia's technology uses the same 2.4GHz band as Wi-Fi does, there's little additional hardware needed to enable existing devices to work with the Cota Tile. "The goal is ultimately, the electronics ... will be just like Wi-Fi, integrated into the device," said Ossia CEO Mario Obeidat.

An IoT Revolution

The ability of the WattUp far-field transmitter and the Cota Tile to transmit data along with power makes them particularly compelling for the Internet of Things. "There's no way you can get to a trillion IoT devices without wireless power," said Jennifer Grenz, Ossia's vice president of marketing, in a company blog post. "Wireless power is going to go hand-in-hand with the evolution of this entire market."

The best example of this is Ossia's Cota Forever Battery. It can be placed in smoke alarms, sparing you from charging them and also allowing for the transfer of data by giving you the Cota Cloud: cloud-based software that allows for power management of various devices.

"Let's say your smoke detector in the kitchen goes off," said Zeine. "The battery will notice that the smoke detector has changed its power-draw pattern and knows that this new pattern is actually the alarm going off. It sends a message to your phone; the Cota Cloud will tell you that, 'Hey, your smoke detector is going off.' This makes dumb devices smart. It brings them to the internet age."

With both Energous and Ossia, the software layer is a key component of what makes them so game-changing, since they allow for intelligent power management. We all have devices that need charging, and when we charge them is up to us. We might charge a phone every day, a wearable every few days, a game controller whenever we use it. With wireless power, the software will make that choice for you based on your usage patterns, lifestyle, and even the time of the day.

Essentially, both the WattUp transmitters and Cota Tile take the burden from the user in deciding what device to charge and when, and take it on themselves. It's as close to a seamless, cable-free existence as you can get.

Cutting One Cord At a Time

In practical terms, this technology is still a few years from implementation for both consumers and enterprises, but potential use cases are already apparent. Both technologies will allow you to effectively cut the cord at home. Phones won't have to be plugged in. Game controllers will charge according to usage. Smoke detectors will never need their batteries changed.

Energous President and CEO Stephen Rizzone Assuming widespread implementation in public spaces such as coffee shops and office buildings, the endless scramble for chargers, cables, and portable batteries could become a thing of the past. "In the future, you would go into a coffee shop. If it has the Cota logo, and you know your phone has Cota, that means you'll get power when you're in the coffee shop," said Obeidat. "That's the kind of standard that we're creating."

But the biggest impact is likely to be felt with low-power-consumption devices that need to be charged (or their batteries changed) infrequently. Rizzone cited the father of Gordon Bell, Energous VP of Marketing as an example of someone who would benefit considerably from Energous' technology, though he doesn't have many connected devices. "My colleague's father wears hearing aids ... What happens is, he's got these little tiny batteries that are the size of a pomegranate seed, and he's gotta put it in there and swap the battery out. It's a nightmare for an older person."

With a mid-field or far-field transmitter, hearing aids would be able to charge throughout the course of an entire day, according to the usage patterns of the individual user.

Of course, devices such as phones and Bluetooth headphones that are more frequently in need of power boosts are also going to see a shift to wireless charging, according to Rizzone: "A lot of devices that you're currently plugging in and plugging in every single day, eventually, that little tiny tip of your micro USB port breaks. It wears out, or you jammed it in the wrong way, salt water gets in there, sweat, etc. That's a failure part or a failure point for that product. So companies are trying to get rid of that."

Radio Beams Everywhere: Are We Safe?

Energous' longer-range transmission technology are certified under FCC Part 18 and Part 15 while Ossia has passed key tests for FCC approval of the Cota Transmitter, and approval is expected shortly. These rules impose RF exposure requirements and limit power transfer levels. RF exposure must be measured for users and bystanders under different wireless configurations. As a general rule, higher RF power is required to operate at a lower frequency.

In the electric vehicle industry, wireless charging has already been widely implemented and can indicate the kind of requirements we might see for electronics devices. According to an article by Joe Bablo, Ken Boyce, and Hai Jiang at Underwriters Laboratory, aside from the normal safety concerns that apply to any electrical transmission (fire, shock), the wireless charging system has to show it can operate in hazardous conditions and detect and shut down when an organic object enters, among other requirements.

Since the higher power Energous and Ossia technologies use the same radio frequencies as Wi-Fi, one would hope that there are no health impacts—but there are surprisingly few scientific answers to this question. The Nation has reported that the wireless industry has tried to limit the amount of research that can be done into the impact of wireless radiation on public health: "Lack of definitive proof that a technology is harmful does not mean the technology is safe, yet the wireless industry has succeeded in selling this logical fallacy to the world. In truth, the safety of wireless technology has been an unsettled question since the industry's earliest days," wrote authors Mark Hertsgaard and Mark Dowie. "The upshot is that, over the past 30 years, billions of people around the world have been subjected to a massive public-health experiment: Use a cell phone today, find out later if it causes cancer or genetic damage."

The Cota beacon signal bounces off walls and objects while avoiding people. Power is delivered along those paths.

According to Ossia, though, Cota technology is safe for people and pets and doesn't interfere with Wi-Fi or face the overheating issues of Qi wireless charging. "2.4GHz and above don't heat metal themselves; they reflect off the metal" Obeidat said.

Energous claims the same. "We're not blasting energy all over," pointed out Rizzone. "We're planting a small, unique pocket of energy just around the device, and it transmits [to] that device."

With FCC certification and no evidence to the contrary, the wireless transmission technologies should be no more dangerous or detrimental than Wi-Fi or any other RF spectrum we're exposed to daily. Perhaps not the most comforting answer, but the only one readily available.

Our Wireless Future

Energous partners with Dialog Semiconductors, a large electronics manufacturer, to produce their transmitters and receivers while Ossia partners with Motherson and Molex, and there have long been rumors about Energous technology ending up in a future Apple product. Neither company manufactures its own hardware, nor do they create consumer products themselves.

Energous is primarily a licensing company, and its products are partner-based. Myant's Skiin fitness- and health-tracking clothes, expected to ship this summer, are the first WattUp-enabled products. And neither Ossia's Cota Forever Battery nor the sizable Cota Tile can be purchased. The company is working on getting the Tile into a smaller, more efficient form-factor.

"The future system that we're working on will achieve the 10 meters and 20 meters," Zeine said. "We're focused at scales where people want to use this, so if they want to put this in the ceiling at Mobile World Congress, you could design a system that could cover 80 or 100 meters. If it's for a home, you don't need a system that's that big."

The first place you might see the larger Tile is in your local Walmart. In this case, you'd have several Cota Tiles providing power to thousands of low-power RFID tags. And the tags, which could show prices, discounts, and bar codes, could all be managed by software. It's not the most exciting implementation, but it's the one most likely to happen soonest.

Wireless Power is Closer

Ultimately, wireless charging technology needs a few more years, a few more product announcements and certifications, and some experimentation by large retailers. So don't throw out your cables and chargers just yet. But the widespread adoption of Qi inductive charging on phones and wearables is strong evidence that both manufacturers and consumers are eager to untether.

Anker's Gao takes a cautiously optimistic view. In one or two years, he said, we should see more and more phones that can charge wirelessly, and he expects that wireless-charging efficiency will have reached the level of wired charging. Gao also said he hopes that in as soon as three to four years, we'll have a "true position-free technology," likely debuting on a wearable device.

Wireless power has always seemed like a work in progress. From Nikola Tesla till now, the promise has always been "the wireless power revolution is coming," yet it never seems to arrive. It remains to be seen if Energous and Ossia can flip the script.

But there's a reason wireless power is an idea that refuses to die. It promises freedom from arranging our lives around all the many power-hungry devices we've accumulated, each one acting as a tether that limits our movement. And that promise is worth waiting for.

Reference[xvi] Article - Wireless Power Is Coming, PC Magazine, Written By Ajay Kumar, June 8, 2018 8:00AM EST. Extracted from the Internet January 7, 2019 from https://www.pcmag.com/article/361628/wireless-power-is-coming

Introduction to Wireless Power (WPT)

We have all grown up in a world where electric wires and cables run everywhere. Wouldn't it be great if we could supply power to our electronic devices anytime we wanted, with no cables and no need for batteries? Wireless electricity sometimes also called Wireless Power Transmission (WPT) may sound like something out of science fiction, but it exists today and looks to be emerging as a big part of our future.

The History of Wireless Power

Scientist Nikola Tesla demonstrated wireless electric lighting more than 100 years ago. Surprisingly little technological progress was made in this area in the ensuing years for whatever reason; some conspiracy theorists claim interference from the large electrical companies of the day is to blame.

Space exploration initiatives of the 1960s triggered the modern wave of research into wireless power. While the long-distance WPT systems that Nikola Tesla dreamed about haven't yet been built, technology advances in short-range WPT began reaching consumers in the 1990s in the form of gadgets like rechargeable electric toothbrushes.

Interest in WPT has exploded in recent years thanks to the popularity of mobile devices. People have grown increasingly frustrated with their phones and tablets running out of charge during the day or having to be plugged in to recharge every night.

Wireless Charging

Short-range wireless charging continues to be by far the most common application of WPT in use today. Traditional WPT relies on a method called inductive coupling but some newer products use magnetic resonance instead. Several different industry efforts continue working to standardize the technology for wireless charging.

A group of companies formed the Wireless Power Consortium in 2008 to promote Qi, a specific inductive coupling technology for wireless charging. Many phones and tablets offer Qi support.

The Power Matters Alliance (PMA) was formed in 2012. PMA directly competes with Qi and has developed its own technical specifications for use of inductive coupling technology.

A third technology for wireless charging named Rezence uses magnetic resonance. A group of companies founded the Alliance for Wireless Power

(A4WP) in 2012 to promote Rezence. In 2014, A4WP and PMA signed agreements to adopt each other's standards.

While many mobile devices support some form of wireless charging, many others do not. Wireless charging will likely gain worldwide adoption over time as the different technical standards mature. Most wireless charging solutions today require the device be located on or very near to the wireless charging unit (such as a mat). Devices must also sometimes be carefully positioned to establish a suitable wireless link.

The Future of Wireless Power

Someday it may be possible to tap into wireless electricity anywhere we are located, maybe even for free, such as if a device could receive power over the same Wi-Fi connections it uses for network data. Both technical and business roadblocks make this vision unlikely to happen anytime soon, however;

Wireless signals over long distances typically suffer from attenuation, interference and other effects that limit their usefulness

Companies that run profitable businesses on the traditional cabled power grids may resist moving to wireless for fear of financial losses.

Concerns over the long-term effects of wireless exposure on the human body persist

Reference[xvii] Article - Introduction to Wireless Power (WPT) Extracted from the Internet January 7, 2019 @ https://www.lifewire.com/introduction-to-wireless-power-electricity-817377 . Written by Bradley Mitchell An MIT graduate who brings years of technical experience to articles on SEO, computers, and wireless networking. Updated June 01, 2018. Bradley is a former writer for Lifewire .

Section Four

Promising Energy Pioneers

The Great Energy Transition: The Years Just Ahead

Alternative Energy

As the global climate continues to deteriorate, consideration of where our civilization is heading becomes more critical with every passing day. The frequency and intensity of hurricanes and cyclones around the world are forcing many to reconsider just how disruptive global warming could become. While many in the U.S. still refuse to accept that the excessive combustion of fossil fuels is the primary cause of global warming, most of the world's leaders are now in agreement that fossil fuels eventually must be phased out.

Another looming problem is how long plentiful supplies of affordable fossil fuels can last. Global oil consumption is now running close to 98 million barrels per day, and the U.S. Department of Energy predicts that global oil consumption will continue to grow for the next 25 years although not quite as fast as the 1.5 million barrels per day annual increase we have seen lately.

A little multiplication shows that the world now is burning through 36 billion barrels of oil each year and soon will be around 40 billion if consumption keeps growing. This, of course, adds up to 400 billion barrels of oil each decade or circa 1 trillion barrels of oil consumption in the next 25 years. Although nobody likes to talk about it, consuming this much oil by the early 2040s is not likely to happen. Either the climate is going to get too bad to continue, or the costs of extracting each additional barrel of oil will get too high in comparison to other alternatives.

The combination of global warming and the depletion of affordable fossil fuels suggests that somewhere ahead, possibly as soon as five to 10 years from now, global civilization is going to be in big trouble unless we can find a viable alternative to fossil fuels as our primary source of energy. The use of solar, wind and maybe even biofuels will grow rapidly and new techniques for storing electrical energy from intermittent sources are likely to come into widespread use in the decade ahead. However, current projections say that renewable fuels will only amount to some 15 percent of our needs 25 years from now. If, as seems likely, consumption of fossil fuels is constrained for one reason or another, the global economy is going to change, perhaps rapidly.

Another and more interesting alternative, however, is whether there are completely new technologies in the offing that could be used as a substitute for fossil fuels — possibly even in time to mitigate the effects of climate change and eventual fossil fuel depletion. Although it may sound far-fetched to many, from time to time, new technologies are discovered and developed which change civilization — think steam, electricity, internal combustion, or electronics.

Currently there are two alternative technologies which seem viable and could be on the market in the next few years. These are Low Energy Nuclear Reactions (LENR) and Brilliant Light Power's (BrLP) SunCell. Both of these technologies have been under development for years, but are based on scientific principles not as yet well understood and controversial. Despite the lack of a well understood scientific basis, both seem to work as has been verified by experiments going back many years. The controversy surrounding these technologies accounts for the lack of publicity in the mainstream media.

Although numerous scientists and organizations around the world are working on LENR, only two currently are saying they are close to launching commercial products. Unfortunately, neither of these entities are very forthcoming about the details of their devices. The Italian, Andrea Rossi, who seems to be as far ahead in developing the LENR technology as anyone, says he is finishing up testing, will be ready for a public demonstration later this year, and start selling heat generated by his devices next year.

BrLP Power and its CEO Randell Mills, however, have always been quite open about the progress they are making and even the difficulties they are encountering. From BrLP's recent presentation we learned that they will not meet their goal of having working prototypes ready for external testing this year. As could be expected from a radically new technology operating at extreme temperatures they have encountered numerous difficulties in turning a laboratory prototype into an automated and reliable device that can be put into commercial service.

At the presentation in Denver, last week Mills released a new schedule which now envisions field testing of his SunCell in the second half of 2018 and a commercial launch in the second half of 2019. This announcement sets the projects commercial launch back by about a year.

Mills, however, outlined several new developments which could turn out to be good news for the SunCell project and speed with which it can be turned into a commercial product. He reports that in the past year, BrLP has overcome a series of "key engineering challenges" in automating the device and that the engineering is far enough along to turn the project over to their partner, Columbia Tech of Boston, MA, to finish the engineering and build the first automated prototypes. Reaching this goal would be a significant achievement, for an automated SunCell should be able to run for extended periods and develop information about its performance and reliability.

The current schedule allows another nine to 15 months to build and test the first prototypes in-house. Hopefully, during this period, there will be a public demonstration of a fully automated SunCell that can produce energy continuously. Such an achievement may be enough to convince the scientific community, the mainstream media, and governments around the world that an alternative to fossil fuels may be available soon.

During the past year, BrLP has rethought its plans and has decided to delay development of the complex concentrated photovoltaic subsystem that was to produce electrical energy directly from the luminescent sphere energized by the hydrino reaction. The company now is planning to develop a "thermal" version of the SunCell which would be capable of delivering 500kW of thermal energy for boiler, hot air or hot water systems. BrLP has contracted with a leading heat exchanger firm, TMI Climate Solutions, to develop the concepts leading to a commercial heat-producing SunCell.

In comparison with concentrated photovoltaics, the design and construction of a thermal energy producing SunCell as opposed to an electricity producing one should be relatively easy. Heat exchangers are an old and well-understood technology even if extracting heat from a plasma glowing at 3,600K is unusual. BrLP likely hopes it can bring a product to market in less time than it takes to develop a concentrated photovoltaic version of the device.

To speed up the electricity-producing version of the SunCell, BrLP has decided to use readily available off-the-shelf photovoltaic cells that should get the device working much more quickly and with fewer development problems. This decision will result in a less efficient electricity-producing version of the SunCell which will only produce 120kW in its first iteration. For a second generation, BrLP now is considering attaching its device to a magnetohydrodynamic device. Such a device would pass the 3,500K plasma produced by the hydrino reaction between two magnets which will produce electricity. Such a device could in theory be very efficient and allow later versions of the SunCell to generate electrical power several times higher than the first generation will be able to do.

While many are frustrated at the seemingly endless delays in bring these new sources of energy to the place where they can produce useful energy, it should be remembered that we are dealing with technologies that are based on science somewhat beyond our knowledge and understanding of nature. Given the potential of Brilliant Light's SunCell technology or various LENR project that are underway, they are well worth waiting for given the current outlook for the world's climate and energy supply.

Reference Article - The Great Energy Transition: The Years Just Ahead. Published September 26, 2017 by Peakoil.com News. Extracted from the Internet January 17, 2019 @ https://peakoil.com/alternative-energy/the-great-energy-transition-the-years-just-ahead[xviii]

A Living Legend, Greater Than Einstein and Tesla Combined

Dr. Randell L. Mills, a Harvard trained medical doctor, who developed the Tokamak ITER fusion reactor generator and the physical theory behind it literally, will overturn the standing theories of physics, dismantle the power grid, put an end to virtually all air pollution, put an end to poverty, put an end to wars over energy resources, open up the entire planet to human habitation, put a serious dent in the ability of politics to control people and save limitless lives in the future due to a lack of power.

Being alive at the end of the twentieth century here on planet Insanity has afforded us a few special privileges in the observance of human history. We've seen the birth of the microprocessor, the creation of the internet, and other technological marvels that have vastly improved the quality of the human condition for billions of people.

Unfortunately, even with all of our modern technological advancements, there's still a huge number of people living in extreme poverty. Most of the poverty can be directly linked to a lack of cheap and abundant energy. According to the IEA, "Modern energy services are crucial to human well-being and to a country's economic development; and yet globally 1.2 billion people are without access to electricity and more than 2.7 billion people are without clean cooking facilities. More than 95% of these people are either in sub-Saharan African or developing Asia, and around 80% are in rural areas."

Even in many places with access to power, it's not cheap enough or abundant enough to lift the population above the poverty line. Access to energy is *the* defining limit on the ability of any given population to pull itself out of poverty by increasing its productive capacity.

All real wealth comes from the ground. Wealth starts as raw materials that are transformed by human intervention into usable goods that improve the quality of the human condition. Extracting and transforming resources from the planet takes a tremendous amount of energy. The cheaper and more abundant the energy resources at our disposal, the cheaper and easier it is to extract and transform resources; which ultimately means cheaper and more abundant goods and services for all of humanity.

For most of human history, the energy to transform resources has come from our own manual labor, the leverage of simple manual tools, and of course – fire. Even today, we live in what could still broadly be categorized as the combustion age. We use combustion to produce the vast majority of energy on the planet, and it's still by far the cheapest means of producing energy yet available to man. We burn coal and oil to power our homes and factories, we burn gasoline to power our cars and planes, we burn crude oil to power our tankers and shipping; it's all a big bloody burning mess.

Numerous wars and insurrections have been fought over the control of energy resources, and all the while, the power grid acts as a control grid that forces massive swaths of people to herd together into compact concrete jungles. The human condition on planet Insanity is so intimately linked with the production of power that we use the same word to describe everything from politics to finance to the performance of an engine – *Power.*

Imagine what the planet would look like if every person had access to a compact generator that was capable of providing enough energy to power a small factory, while requiring no fuel, producing no pollution, requiring virtually no maintenance for decades, and being portable enough to fit in a small vehicle. It's hard to fathom because the potential for invention and mobility becomes nearly limitless. People could live comfortably in the middle of the ocean or in the middle of the Alaskan wilderness, it wouldn't make a bit of difference. Homes could be flying recreational vehicles that never have to land. Cars could be airplanes and airplanes could be transformed into space vehicles. With such an invention, we could even colonize the inner planets of the solar system.

With such an invention, the power grid becomes obsolete. Humans become untethered to the entire utility grid, water and sewage included. Water could be pulled from the air or desalinated from the oceans. Trash and sewage could be incinerated on demand. Unlimited personal power brings with it virtually limitless potential. Unlimited personal power brings with it virtually limitless freedom.

I'm telling you now that such an invention exists, and in the next year or so, it's going to change the face of humanity forever. This is the man who will do it;

Dr. Randell L. Mills serves as the Chairman, President, and Chief Executive Officer at Brilliant Light Power, Inc. He served as the Chairman, Chief Executive Officer and President of BlackLight Power Inc. since 1991. Dr. Mills founded BlackLight Power, Inc. Dr. Mills has authored eight books, participated in over 50 presentations at professional meetings, and authored and co-authored over 100 papers regarding the field of energy technology that have been published in or submitted to peer-reviewed journals of international repute. Dr. Mills has received patents or filed patent applications in the following areas: (1) Millsian computational chemical design technology based on Classical Physics (CP), a revolutionary approach to solving atomic and molecular structures; (2) magnetic resonance imaging; (3) Mossbauer cancer therapy (Nature, Hyperfine Interactions); (4) Luminide class of drug delivery molecules; (5) genomic sequencing method, and (6) artificial intelligence. A thorough description of the Company's technology and Dr. Mills' underlying atomic theory is published in a book entitled The Grand Unified Theory of Classical Physics. Dr. Mills holds a Bachelor of Arts Degree in Chemistry, summa cum laude and Phi Beta Kappa, from Franklin & Marshall College in 1982, and a Doctor of Medicine Degree from Harvard Medical School in 1986. Following a year of graduate work in electrical engineering at the Massachusetts Institute of Technology, Dr. Mills began his research in the field of energy technology.

Dr. Randell L. Mills, a Harvard trained medical doctor, who developed this generator and the physical theory behind it, will be remembered throughout the remainder of human history as the single greatest being to ever have lived. A billion trillion years from now, people will still be discussing his theories and his invention. Mills invention is no less transformative than the discovery of fire. It's a far greater achievement than the combustion engine or the telephone.

In one fell swoop, Mills will have overturned the standing theories of physics, dismantled the power grid, put an end to virtually all air pollution, put an end to poverty, put an end to wars over energy resources, opened up the entire planet to human habitation, put a serious dent in the ability of politics to control people, and saved limitless lives in the future due to a lack of power.

To be alive at the same time as Mills is a cosmic treat. The man is a living legend that has yet to be recognized. Like all great transformative scientists before him, academia has viciously attacked him, his theories and his company, on both a personal and professional level. Mills is truly the modern day Galileo. Like Mills, Galileo was viciously attacked by academia for his ideas, even though they are basis for all modern science today, to the point where Galileo spent the last days of his life under house arrest for scientific heresy.

Here's what mainstream science has had to say about Mills and his work.....

I feel compelled to copy and paste this here for posterity, because once Mills' generator hits the market, these comments are sure to be flushed down the memory hole. Brilliant Light Power (formerly Black Light Power) is the name of Mills' company that's producing the generator and doing the research. The following is taken from Brilliant Light Power's wiki page (https://en.wikipedia.org/wiki/Brilliant_Light_Power), which Brilliant Light obviously has no control over (think about that for a minute, a company that's prohibited from editing its own wiki page highlights just how terrified academics are of Mills and his theories):

Robert L. Park

Since 1999 Robert L. Park, emeritus professor of physics at the University of Maryland and a notable skeptic, has been particularly critical of BLP. In 2008 Park wrote:

"BlackLight Power (BLP), founded 17 years ago as HydroCatalysis, announced last week that the company had successfully tested a prototype power system that would generate 50 KW of thermal power. BLP anticipates delivery of the new power system in 12 to 18 months. The BLP process, discovered by Randy Mills, is said to coax hydrogen atoms into a "state below the ground state", called the "hydrino." There is no independent scientific confirmation of the hydrino, and BLP has a patent problem. So they have nothing to sell but bull shit. The company is therefore dependent on investors with deep pockets and shallow brains." – Park

Steven Chu

In 1999, Steven Chu (https://en.wikipedia.org/wiki/Steven_Chu), Nobel Laureate in Physics in 1997, said "it's extremely unlikely that this is real, and I feel sorry for the funders, the people who are backing this".

Wolfgang Ketterle

Wolfgang Ketterle, (https://en.wikipedia.org/wiki/Wolfgang_Ketterle) a professor of physics at MIT, said BLP's claims are "nonsense" and that "there is no state of hydrogen lower than the ground state".

Michio Kaku

Michio Kaku (https://en.wikipedia.org/wiki/Michio_Kaku) , a theoretical physicist based at City University of New York, adds that *"the only law that this business with Mills is proving is that a fool and his money are easily parted."* and that *"There's a sucker born every minute."*

Peter Zimmerman

While Peter Zimmerman (https://en.wikipedia.org/wiki/Peter_Zimmerman) was chief arms-control scientist at the State Department, he stated that his department and the Patent Office "have fought back with success" against "pseudoscientists" and he railed against, among other things, the inventors of "hydrinos."

Forbes attacked him, saying, "This case has many things in common with other failed new energy technologies, including a charismatic leader who generates lots of press, claims of academic validation and criticism of expert skepticism, ample fundraising, and frequent reports of interest from potential buyers. What you don't find are actual ongoing operation of a power plant outside of demonstration models controlled by the promoter."

The IEEE attacked him, calling him a loser. And of course, countless internet trolls have attacked him and his company over all the scientific forums scattered across the internet. People have been banned simply for posting about Mills and his theories on sites like physicsforums.com and the Reddit science boards. You can't even discuss his theories on the major science forums without risking an immediate ban.

Academics typically don't even bother commenting on theories that have no scientific merit, so when you see this level of vitriol over Mills' theories, it stands to reason it's being driven by fear, not by a genuine motivation to seek the truth.

I've stated numerous times on this site over the past decade that modern physics is fundamentally flawed. It is not tied to reality. Mills theory takes physics back to its roots. It does away with all the imaginary particles and nonsense that make up the basis of quantum mechanics. Mills theory is based on closed form classical physics, which treats the electron as singular entity that can be modeled and calculated in precise detail. Mills theory allows for the precise calculation of molecular shapes and bond strengths, and will transform the way material sciences are conducted in the future.

Taking away all the imaginary particles and physics demystifies quantum mechanics. With Mills' theory, theoretical physicists are no longer needed, and their positions as the high priests of science are directly threatened. It's little wonder they hate Mills so much. Eventually the theoretical people will be shunted to the side like most religions have been today. Make no mistake, theoretical physics is a religion, not a science. It's incredibly hard to change people's' religious view points since they aren't based on evidence or logical reasoning.

To put Mills work in perspective, the Tokamak ITER fusion reactor has had 14 billion dollars of public funding thrown at it with no results to show for it. Further, to quote wiki, "the ITER is not designed to produce electricity, but made as a proof of concept reactor for the later DEMO project." So 14 billion just to build a proof-of-concept.

Mills has taken less than 100 million of *private* investment capital over less time than the ITER to go from theory, to concept, to design, and now on to actual power production, with less than a tenth of the resources. Further, the ITER will obviously be abandoned after Mills generator goes into production, resulting in 14 billion dollars of public money being flushed down the toilet to feed the pocket books of academics and state contractors. The academics and contractors are the real scam artists in this melodrama. Perhaps that's another reason they attack Mills with such ferocity, they are having a bit of Freudian projection going on within themselves.

So let's talk a little bit about the generator itself. In simple terms, it works by reacting a very tiny amount of hydrogen with a silver catalyst and then exposing that reaction to an electrical discharge. When the hydrogen, silver and electrical discharge combine, the hydrogen atoms shrink, releasing a tremendous amount of energy in the process. Using Mills process, a gallon of water has the same energy potential of around 2000 gallons of gasoline. The amount of hydrogen needed is so small that the generator can pull the hydrogen it needs from water in the surrounding air. This enormous release of energy is then used to basically form the filament of a carbon light bulb.

The carbon light bulb glows so brightly from the heat that it reaches the equivalent brightness of over 2000 suns, with the potential to go even higher in the future. The light bulb is surrounded by an array of special solar cells called concentrator solar cells that are able to harness the energy from this enormously bright light source. These special solar cells already exist and are in use in solar farms around the world today.

Mills is projecting his initial generators to produce around 150 kw of power, with the potential for much higher amounts in the future, and the ability to chain multiple units together to provide huge amounts of power. He projects around 600 or so dollars per unit in materials costs to produce the unit, with the unit weighing far less than a conventional gasoline engine at around the same size. He also says production of the units can be scaled up to near infinite capacity, allowing a tremendous number of these units to be produced for next to nothing, using the same format as current LED production uses today.

Mills already has the "light bulb" portion of this generator fully prototyped and working. Obviously that's the hard part since the rest of the parts for the generator already exist from off-the-shelf suppliers. Mills is projecting production units to be ready for delivery by the second half of 2017, with a fully prototyped working generator ready for field testing by the first half of 2017. Once Mills rolls out that first generator, be prepared for a monumental media shit storm to hit the public.

You can read the latest press release from Brilliant Light Power here - (https://www.businesswire.com/news/home/20160711005351/en/Brilliant-Light-Power-Announces-Validation-Generation-Million), and watch the "light bulb" in action below.

I've been following BLP closely for nearly a decade now. I'm convinced Mills is the real deal. Contrary to what Forbes claims, numerous independent labs have verified his claims. He's obtained funding from some of the wealthiest people on the planet, and he's got a highly skilled team of engineers backing him. He doesn't need any more investment capital, so he's got no reason to lie. Several university labs and independent investigators have ***replicated and validated*** Mills work:

- **Rowan University's report**

https://brilliantlightpower.com/wp-content/uploads/pdf/Chary%20Redacted%20Report.pdf

- **Rowan University's report II**

https://brilliantlightpower.com/wp-content/uploads/pdf/Jansson%20Redacted%20Report.pdf

- **UNC Asheville's report**

https://brilliantlightpower.com/wp-content/uploads/pdf/Dr.%20Booker%20Redacted%20Report.pdf

- **University of Illinois report**

https://brilliantlightpower.com/wp-content/uploads/papers/GlumacReportwithGraphics2014.pdf

It's also worth taking a look at Brilliant Light's board of directors. These people don't strike me as the type of people who are willing to throw their lives away over fraud charges:

H. McIntyre Gardner's profile from Bloomberg:

https://www.bloomberg.com/research/stocks/people/person.asp?personId=1364252&privcapId=831552

- Chairman of Spirit Airlines
- Chairman of Merrill Lynch Bank USA
- Chairman of Merrill Lynch Life Insurance Company and Transamerica Advisors Life Insurance Company of New York
- Director of Isola Group, SARL
- President of the Board of Trustees at the Princeton Charter School

James E. Canty J.D.'s quick profile

- President and General Counsel at Clough Capital Partners
- Associate General Counsel, Corporate Secretary, and Director of Investor Relations at Converse Inc.
- Corporate and Securities Lawyer at Goldstein & Manello
- Certified Public Accountant at KPMG

John J. Gillen's profile from Bloomberg:

https://www.bloomberg.com/research/stocks/private/person.asp?personId=13397084&privcapId=61802124&previousCapId=224710025&previousTitle=Bright%20Plain%20Renewable%20Energy,%20LLC

- Global Director of Finance at Towers Perrin
- Chief Financial Officer and Senior Vice President of Dayton Power and Light Co.
- Partner, PricewaterhouseCoopers
- Member of the Office of the Chief Accountant of the Federal Energy Regulatory Commission

Jeremy Hux:

- Managing director and global head of clean technology investment banking for Credit Suisse
- President of HCP Advisors
- Head of West Coast Clean Technology for Morgan Stanley

Barbara A. Nabial's profile from Bloomberg:

https://www.bloomberg.com/research/stocks/private/person.asp?personId=62447006&privcapId=61802124&previousCapId=61802124&previousTitle=BlackLight%20Power%20Inc.

- Senior Audit Manager with Deloitte & Touche

And let's not forget Mills, who has a Bachelor of Arts Degree in Chemistry, summa cum laude and Phi Beta Kappa from Franklin & Marshall College, and a Doctor of Medicine Degree from Harvard Medical School. He also did graduate engineering coursework at MIT.

At this point, it doesn't matter what anyone may say about him. He's going to market, and he's going to change the world.

Update 10/28/16:

Pretty crazy update from BLP today.

They refactored the SunCell to use dual liquid silver metal injectors for the electrodes, which solved the problem of the reaction being so strong that it was blowing holes through solid tungsten electrodes in just a few seconds.

What's pretty crazy is they are reporting that they can actually shut the power off to the electrodes and the EM pump; yet the reaction will continue in a self-sustaining manner. You can actually see this take place in the video.

BLP states, "It is predicted that with sufficient silver vapor pressure, the power will persist by a self-sustaining hydrino reaction. Moreover, as designed, when the cell is operated as a silver boiler, the EM pump power may be terminated as well such that the parasitic load is zero except for the power consumed in electrolysis of water to provide the hydrogen fuel. In fact, the power did persist at the same level for long duration after the ignition power was terminated (See sections of the run indicated by the ignition-off video captions)."

Update 11/2/2016:

I just read through the latest BrLP presentation event notes and found their current planned development and production timeline. They are projecting the commercial units to produce 100 kW of power.

BrLP has contracted with Columbia Tech to have them do the prototype and manufacturing workups. CT's Chief Technology Officer is projecting initial field testing of prototype units to take place in 6 months, with pilot commercial units available 6 months after initial field testing completes. Phase 1A design began on November 1st.

Masimo Semiconductor is responsible for creating the commercial photovoltaic converter comprising a denser receiver array of concentrator multijunction photovoltaic cells that will wrap around the black-body light source. Masimo's Head of Business Development is projecting the photovoltaic array to be ready for the prototype field tests on time.

Looking at the slide notes from Masimo, they are projecting 15.5kW output for the first prototype, and 28.5kW for the second prototype.

Here's Masimo's published timeline:

-Week 1: July 27, 2016.... Masimo Semi currently under TPV Cell contract and drawings

-Week 10: Oct 19, 2016.....Masimo Semi currently under gDRA prototype development contract

-Week 14: Nov 4, 2016.....1-Junction Cells on GaAs & InP complete

-Week 16: Nov 18, 2016....gDRA Prototype 3-D model complete

-Week 24: Jan 13, 2017....gDRA Prototype (1-J) Single TRU complete

-Week 28: Feb 17, 2017.... 2-Junction Cells on GaAs or InP complete (requires contract)

-Week 40: May 7, 2017....gDRA Prototype (2-Junction Cells) complete (requires contract)

So looking at the notes, Jan 13, 2017 is a big day because that's the day the first geodesic Dense Receiver Array will be completed (gDRA). The gDRA is the photovoltaic dome that wraps around the black body light source. Once that's done, we could see a self-sustaining prototype generator produce power for the first time not too long after that. So perhaps a fully working prototype will be wheeled out for a public demo sometime near the end of January or in February.

Update 11/4/2016:

Brilliant Light Power (BrLP) released video of their October 26th Industry Day, which I just discussed above.

Update 11/2/2016:

I just read through the latest BrLP presentation event notes and found their current planned development and production timeline. They are projecting the commercial units to produce 100 kW of power.

BrLP has contracted with **Columbia Tech** to have them do the prototype and manufacturing workups. CT's Chief Technology Officer is projecting initial field testing of prototype units to take place in 6 months, with pilot commercial units available 6 months after initial field testing completes. Phase 1A design began on November 1st.

Masimo Semiconductor is responsible for creating the commercial photovoltaic converter comprising a denser receiver array of concentrator multijunction photovoltaic cells that will wrap around the black-body light source. Masimo's Head of Business Development is projecting the photovoltaic array to be ready for the prototype field tests on time.

Looking at the slide notes (https://www.brilliantlightpower.com/wp-content/uploads/presentations/Masimo-BrLPIndustryDay-102616.pdf) from Masimo, they are projecting 15.5kW output for the first prototype, and 28.5kW for the second prototype.

Here's Masimo's published timeline:

-Week 1: July 27, 2016.... Masimo Semi currently under TPV Cell contract and drawings
-Week 10: Oct 19, 2016.....Masimo Semi currently under gDRA prototype development contract
-Week 14: Nov 4, 2016.....1-Junction Cells on GaAs & InP complete
-Week 16: Nov 18, 2016....gDRA Prototype 3-D model complete
-Week 24: Jan 13, 2017....gDRA Prototype (1-J) Single TRU complete
-Week 28: Feb 17, 2017.... 2-Junction Cells on GaAs or InP complete

(requires contract)
-Week 40: May 7, 2017....gDRA Prototype (2-Junction Cells) complete (requires contract)

So looking at the notes, Jan 13, 2017 is a big day because that's the day the first geodesic Dense Receiver Array will be completed (gDRA). The gDRA is the photovoltaic dome that wraps around the black body light source. Once that's done, we could see a self-sustaining prototype generator produce power for the first time not too long after that. So perhaps a fully working prototype will be wheeled out for a public demo sometime near the end of January or in February.

Update 11/4/2016:

Brilliant Light Power (BrLP) released video of their October 26th *Industry Day*, which I just discussed above.

Reference[xix] Article - Randell L. Mills – A Living Legend, Greater Than Einstein and Tesla Combined Article posted in the Libertarian News. Written by Michael Suede July 12, 2016. Extracted from the Internet January 17, 2019 @ https://www.libertariannews.org/2016/07/12/randell-l-mills-a-living-legend-greater-than-einstein-and-tesla-combined/

Futurist Dr. Randell Mills Talks SunCell, Off-Grid Power, And The Future Of Job Creation

Jobs, Musk...Mills? Every now and then, a revolutionary thinker imagines a future the rest of us can't, or in the case of Randell Mills, imagines technology that defies the laws of quantum mechanics. Initially mocked, Jobs retains a godlike status, even posthumously. And Musk, well, he's proved skeptics wrong for years, and yet his talk of Hyperloop Pods traveling at hundreds of miles per hour under the streets of L.A. seems like fantasy to many.

If there is one thing I've learned from working for and alongside hundreds of entrepreneurs over the last two decades it is this: pay attention to big thinkers whose ideas presently seem unimaginable, especially when these thinkers are determined to transform the world.

Give me the chance to connect with them personally, and I'm all in.

Not without his own skeptics, I recently had the chance to sit down with Mills and hear about his latest invention, the Sun Cell, which promises to bring clean and cheap energy to the world. As we chatted, I found myself imagining the possibilities and the potential. Lifting millions from poverty? Check. Tackling climate change. Check again.

Mills says the SunCell works by generating electricity with hydrogen being converted to dark matter by using water in the air, and the reaction packs 200 times the energy of burning conventional gasoline. Sound too good to be true? Well, Mills is betting SunCell will soon be commercialized, and strategic investors are backing that bet. Brilliant Light Power (which Mills founded in 1991), has raised $120M to date and has recently completed a $20M funding round.

Imagine living in a world where the grid does not exist and where everyone has access to power, no matter who you are or what part of the world you live. That would be something else.

Rebekah Iliff: Do you consider yourself an inventor, an innovator, an entrepreneur, a social entrepreneur, or a futurist?

Randell Mills: All of the above. To do what I do, you not only need to be an inventor and a theorist but also have a firm comprehension of how to make things work in practice. When your goal is nothing less than delivering a power source greater than fire, your only choice is to be multi-faceted. Otherwise, you're just making incremental improvements, not holistic leaps forward. High-energy dark-matter power as a business seems totally impractical. It is barely fathomable, so you've got to tackle theory, innovation, invention, and practical business simultaneously.

RI: You are a trained medical doctor. Why did you choose to focus on energy?

RM: If you look broadly at science and technology, one thing is ultimately connected to another, and energy just naturally called to me. In fact, I've invented in a number of different areas, including hydrogen energy technology, computational chemical design, magnetic resonance imaging, drug delivery, artificial intelligence and more. But there's really no better opportunity to work on something that could be so profoundly disruptive.

RI: Explain the SunCell for dummies.

RM: It's a massive lightbulb that is on 24/7 and produces cheap and clean energy from the hydrogen atoms of water. It's lit by a reaction between of hydrogen of water molecules to dark matter using the humidity in the air as the water source. It's over 1000 times as powerful as high-octane gasoline, and the power is directly converted to electricity using photovoltaic cells.

RI: What does the world look like when you've commercialized the SunCell?

RM: Everything will be powered by the SunCell. Solar, wind, bio fuels, and nuclear will all be replaced. The grid will be unnecessary. Utilities will be unnecessary. There would be no pollution and limited energy regulation. As the SunCell is fully autonomous, energy delivery becomes impervious to disruption from war, terror, and natural disaster. Importantly, underdeveloped countries will have the same potential lifestyle and productivity as the developed world. Each SunCell could also serve as a self-powered, autonomous node in a mesh network that could replace the Internet.

RI: How would it impact jobs?

RM: Jobs are created by wealth. If you have something that encourages productivity, then there will be jobs. The SunCell encourages productivity by leveling and equally distributing the energy playing field for virtually anyone, anywhere. There is literature about GDP and energy dependency, and it's very revealing in terms of how dependent we are on power.

RI: I know a lot of folks at your level would let ego get in the way. How do you stay grounded?

RM: My background is helpful. I didn't come from the Ivy League. Most people in my life were not Harvard or MIT graduates. They weren't captains of industry. They were honest and worked hard. I grew up on a farm in Cochranville, Pennsylvania, where life was very challenging and humbling. You earned an appreciation for dangerous equipment at an early age. I've interacted with all types of people from every stature in life, and I've always maintained that simple farmer's perspective.

RI: What is your PR strategy around this? How do you plan on shifting the public opinion enough to override business as usual?

RM: We have a multi-pronged approach. We're introducing state-of-the-art theory, science, and technology that astonishes experts with quantifiable and verifiable results in multiple scientific and technological fields, and we are building a machine with a story behind it that blows everyone else away. Within a couple months we should be ready to show its commercial potential.

RI: Any parting thoughts you'd like to share?

RM: I think the next "age" is an age where mankind has a manual of the universe and knows exactly how the universe works. We then begin to create previously unimaginable inventions by applying this manual and the newly discovered laws that come along with it. The next big future is the physical age.

"The ramifications of bringing cheap and clean energy to the world are transformational. Ultimately, I'm working to change our perception of the universe. So instead of having tunnel vision into a single aspect of quantum theory, I had to become a cross disciplinarian."

"Practically speaking, I think that beyond SunCell, visionaries like Elon Musk will introduce entirely new technologies that people haven't even conceived of. The pace will only increase in the coming years, as we will see similar types of disruptive and enabling technologies come to market." - Dr. Randell Mills

Reference[xx] Article - Futurist Dr. Randell Mills Talks SunCell, Off-Grid Power, And The Future Of Job Creation. Written by Rebekah Iliff, contributing writer to the HUFFPOST and is currently the Chief Strategy Officer for AirPR, a PR technology company. Posted June 01, 2017. Extracted from the Internet January 17, 2019 @ https://www.huffingtonpost.com/entry/futurist-dr-randell-mills-talks-suncell-off-grid_us_592ec431e4b07c4c73138706

Finally! Independent Testing Of Rossi's E-Cat Cold Fusion Device: Maybe The World Will Change After All

Back in October 2011 when Mark Gibbs first wrote about Italian engineer, Andrea Rossi, and his E-Cat project, a device that produces heat through a process called a Low Energy Nuclear Reaction (LENR).

Very briefly, LENR, otherwise called cold fusion, is a technique that generates energy through low temperature (far lower than hot fusion temperatures which are in the range of tens of thousands of degrees) reactions that are not chemical. Most importantly, LENR is, theoretically, much safer, much simpler, and many orders of magnitude cheaper than hot fusion. Rather than explaining LENR in detail here please see my original posting for a more complete explanation.

My next post on this topic was here on Forbes a few days later and, as the labyrinthine and occasionally ridiculous saga developed, I tried to sort fact from fiction in a series of posts (see the list at the end of this posting) which covered everything from unconvincing demos, through an Australian businessman offering Rossi $1 million to show independently tested proof, to other players in the LENR market showing interesting results.

I haven't posted about Rossi and his E-Cat since last August simply because there wasn't much to report other than more of Rossi's unsupported and infuriating claims that included building large-scale automated factories to churn out millions of E-Cats (the factories still have no sign of actually existing) through to unsubstantiated performance claims that sounded far too good to be true.

What everyone wanted was something that Rossi has been promising was about to happen for months: An independent test by third parties who were credible. This report was delayed several times to the point where many were wondering whether it was all nothing more than what we have come to see as Rossi's usual "jam tomorrow" promises. But much to my, and I suspect many other people's surprise, a report by credible, independent third parties is exactly what we got.

Published on May 16, the paper titled "Indication of anomalous heat energy production in a reactor device" would appear to deliver what we wanted.

The paper was authored by Giuseppe Levi of Bologna University, Bologna, Italy; Evelyn Foschi, Bologna, Italy; Torbjörn Hartman, Bo Höistad, Roland Pettersson and Lars Tegnér of Uppsala University, Uppsala, Sweden; and Hanno Essén, of the Royal Institute of Technology, Stockholm, Sweden.

While some of these people have previously been public in their support of Rossi and the E-Cat they are all serious academics with reputations to lose and the paper is detailed and thorough.

The actual test reactor, called the E-Cat HT, was described by the testers as:

... a high temperature development of the original apparatus which has also undergone many construction changes in the last two years – is the latest product manufactured by Leonardo Corporation: it is a device allegedly capable of producing heat from some type of reaction the origin of which is unknown. They described the E-Cat HT as:

... a cylinder having a silicon nitride ceramic outer shell, 33 cm in length, and 10 cm in diameter. A second cylinder made of a different ceramic material (corundum) was located within the shell, and housed three delta-connected spiral-wire resistor coils. Resistors were laid out horizontally, parallel to and equidistant from the cylinder axis, and were as long as the cylinder itself. They were fed by a TRIAC power regulator device which interrupted each phase periodically, in order to modulate power input with an industrial trade secret waveform. This procedure, needed to properly activate the E-Cat HT charge, had no bearing whatsoever on the power consumption of the device, which remained constant throughout the test. The most important element of the E-Cat HT was lodged inside the structure. It consisted of an AISI 310 steel cylinder, 3 mm thick and 33 mm in diameter, housing the powder charges. Two AISI 316 steel cone-shaped caps were hot-hammered in the cylinder, sealing it hermetically.

There were two test runs of the E-Cat HT (the emphasis is mine):

The present report describes the results obtained from evaluations of the operation of the E-Cat HT in two test runs. The first test experiment, lasting 96 hours (from Dec. 13th, 2012, to Dec. 17th, 2012), was carried out by the two first authors of this paper, Levi and Foschi, while the second experiment, lasting for 116 hours (from March 18th 2013, to March 23rd 2013), was carried out by all authors.

The authors also note various assumptions they made about the test and that they weren't in control of all of the aspects of the process, but they apparently didn't consider any of these to be egregious enough to be showstoppers.

And now, the big reveal ... the authors' conclusions are (again, the emphasis is mine):

... if we consider the whole volume of the reactor core and the most conservative figures on energy production, we still get a value of (7.93 ± 0.8) 102 10^2 MJ/Liter that is one order of magnitude higher than any conventional source.

To put that in perspective, the following graph plots the peak power of various energy sources against their specific energy (energy per unit mass). As you can see, gasoline is way out in front in terms of how much energy is available and how much power can be delivered but if this paper is correct, you can make that "gasoline was way out in front" because, as can be seen, the E-Cat has roughly four orders of magnitude more specific energy and three orders of magnitude greater peak power than gasoline!

While a few commentators have raised criticisms concerning how the measurements were made and sources of error others have argued that the energy produced is so significant even knocking off an order of magnitude on either axis still portrays a process with insanely valuable output.

This is not, of course, the last word or even one anywhere near the end of this story but unless this is one of the most elaborate hoaxes in scientific history it looks like the world may well be about to change. How quick will depend solely on Rossi.

Reference Article - Finally! Independent Testing Of Rossi's E-Cat Cold Fusion Device: Maybe The World Will Change After All Written by Mark Gibbs 2013-05-20 for Forbes and Posted: 2013-05-28 10:02:12. Extracted from the Internet 2 January 2019 from https://www.forbes.com/sites/markgibbs/2013/05/20/finally-independent-testing-of-rossis-e-cat-cold-fusion-device-maybe-the-world-will-change-after-all/#2b39775b24dd [xxi]

Signs of Hope - Stan Ovshinsky

I came to Detroit and its environs, the seat of America's glorious industrial past, to see if I could get a glimpse of the future. Is the economic, social and physical deterioration that has caused so much misery in the Motor City a sign of what's in store for larger and larger segments of the United States?

Or are there new industries waiting in the wings — some of them right here in the Detroit metropolitan area — with new jobs and bright new prospects for whole new generations of American dreamers?

I found real reason to hope when a gentleman named Stan Ovshinsky took me on a tour of a remarkably quiet and pristine manufacturing plant in Auburn Hills, which is about 30 miles north of Detroit and is home to Chrysler's headquarters. What is being produced in the plant is potentially revolutionary. A machine about the length of a football field runs 24 hours a day, seven days a week, turning out mile after mile after mile of thin, flexible solar energy material, from which solar panels can be sliced and shaped.

You want new industry in the United States, with astonishing technological advances, new mass production techniques and jobs, jobs, jobs? Try energy.

Mr. Ovshinsky knows as much or more about the development and production of alternative energy as anyone on the planet. He developed the technology and designed the production method that made it possible to produce solar material "by the mile." When he proposed the idea years ago, based on the science of amorphous materials, which he invented, he was ridiculed.

But the thin-film photovoltaic solar panel was just one of his revolutionary ideas. He invented the nickel metal hydride battery that is in virtually all hybrid vehicles on the road today. And when I pulled into the parking lot outside his office in Bloomfield Hills, he promptly installed me in the driver's seat of a hydrogen hybrid prototype — a car in which the gasoline tank had been replaced with a safe solid-state hydrogen storage system invented by Mr. Ovshinsky.

Within minutes, I was driving along a highway in a car that produced zero pollution. No carbon footprint whatsoever. How's that for a wave of the future?

The point is that these (and many more) brilliant, innovative technologies are here. They are real, tangible. They exist. What's needed now is the will to develop policies that will vastly expand these advances and radically reduce their costs. The United States should be leading the world in the creation of whole new energy technologies and industries, instead of allowing the forces of the old carbon-based industries — coal, oil, gasoline-powered vehicles — to stand obstinately in the way of real progress.

"Now," Mr. Ovshinsky told me, "is when we have to build the new industries of the future." He has always been driven by the desire to use science and technology to solve the real-world problems of real people, and that has meant creating employment and stopping the pollution of the planet. He and his late wife, Iris, formed a company (to become known as Energy Conversion Devices) in Detroit in 1960 with the idea of using their considerable talents, as he put it, "to do good, to change the world."

After nearly a half-century of revolutionary innovations with the company, Mr. Ovshinsky retired two years ago to focus his attention on the difficult and time-consuming effort to make solar energy economically competitive with coal and oil. "I know solar energy can't live up to its possibilities unless it's a hell of a lot cheaper," he said.

He believes he has assembled a team that, with sustained, intense work under his direction — and if sufficient funding can be secured — will bring the price of solar power below that of coal and oil within a few years.

What's weird is that this man, with such a stellar track record of innovation on products and processes crucial to the economic and environmental health of the U.S., gets such little attention and so little support from American policy makers. In addition to his work with batteries, photovoltaics and hydrogen fuel cells, his inventions have helped open the door to flat-screen televisions, new forms of computer memory and on and on.

So when Stan Ovshinsky tells us that we should be putting our chips on hybrid and electric vehicles, and that solar and hydrogen power can be the cornerstone of an industrial renaissance in the U.S. as well as a cleaner planet, we should be listening very, very closely.

As oil defined the 20th century, new forms of energy will define the 21st. The U.S. has the opportunity, the intellectual resources and the expertise to lead the world in the development of clean energy. What we've lacked so far has been the courage, the will, to make it happen.

Reference[xxii] Article - Signs of Hope, by Bob Herbert for The New York Times November 23, 2009. Extracted from the Internet January 15, 2019 from https://www.nytimes.com/2009/11/24/opinion/24herbert.html

Scalar Energy Tom Bearden - A Completely New World Is Possible

The greatest scientific discovery in the history of the world has happened over the last few decades and it has gone largely unnoticed because of the great secrecy with which it has been held by all who know of it. It is the discovery of a completely new kind of electromagnetic waves which exist only in the vacuum of empty space, the empty space between the atoms of our bodies as well as the empty space we see in sky at night. All of empty space.

These waves constitute a kind of ocean of infinite energy, and it has now been discovered that this abundant energy can be coaxed to pour into our 3-dimensional world from their 4-dimensional realm, to be used to do work, provide electricity, power all transport, and even heal the body of almost all disease. This is the new world of scalar electromagnetics, the zero-point energy, the energy of the absolute nothingness which existed before the world began.

To understand the incredible implications of this phenomenal discovery is to realize that a completely New World is possible for mankind, a future never before dreamed possible. It could be a world of abundance everywhere, of endless free energy, pollution-free, and a world of cheap universal healthcare, in which actual healing takes place instead of mere treatment, and for most ailments, without the load of chemicals of the pharmaceutical industry. It could be a world in which a huge portion of this world's massive suffering could be relieved.

These new waves of energy are called "longitudinal" EM (electromagnetism) to distinguish them from "transverse" EM, the kind we are familiar with in our daily life, which power our cell phones and pagers, television and radio broadcasts, microwave ovens, and so on. Rather than modulating in 3-dimensions they are modulating in the direction they are going, accordion-like, that is, along the axis of time, the 4th dimension. What this means is still a mystery to me, but one important facet of this new discovery in the new field of "scalar electromagnetics" is the discovery that time itself is compressed energy, compressed by the factor of the speed-of-light-squared. Does that sound familiar? It brings to mind Einstein's E = mc2. Matter itself is also compressed energy, and compressed by the same factor as scalar energy is compressed in time. So the discovery of scalar electromagnetics is that we can tap and use the immense energy existing in the ocean of time.

This amazing discovery is announced and described by foremost U.S. scalar scientist Tom Bearden on his website Cheniere.org. I spent about six months poring over the papers and briefings there, despite being a layman and not understanding the high physics and mathematical formulas of it. But I could glean enough to know that, if Bearden is correct, this discovery of the zero-point energy was in fact and without doubt the greatest discovery in human history.

This discovery will completely change our understanding of physical and "non-physical" reality. And it could lead to heaven, and it could lead to hell, because the incredible powers of these new longitudinal waves can, and already HAVE BEEN harnessed in the creation of the most stupendous weapons the world has ever seen. The least of them, the longitudinal interferometer or "Tesla howitzer," could bring down, in a single shot, a huge building the size, say, the World Trade Center towers, and is being used daily to engineer and change the weather over North America. The Russians are probably the most advanced in the development of these kinds of weapons, but the Chinese are catching up as well. Once you have the know-how building a Tesla howitzer is relatively cheap, the key is in the know-how, knowing how to manipulate, modulate and intersect the longitudinal waves at-a-distance. You can be anywhere in the world and deliver a shot to anywhere else in the world with a Tesla howitzer. All nations now developing these new electromagnetic weapons do so in great secrecy, so mankind at large remains unaware of the danger into which it has fallen.

It should be noted that this discovery of the longitudinal waves of the vacuum of space is not so new as it might seem. Nikola Tesla is generally considered the father of scalar electromagnetics. Tesla's name for this energy, also called "scalar energy" or "zero-point" energy was "Radiant Energy." Tesla worked in the first half of the last century and was clearly a genius of the highest order. Unfortunately Tesla was quite paranoid about others stealing his inventions and so, with his photographic memory, kept most of what he knew in his head rather than write it down. But he made a free energy device called the "Tesla turbine," and even converted an automobile to run on the "radiant energy." At Tesla's death the government swooped in immediately and confiscated what writings and papers there were. So, many of Tesla's secrets were lost to humanity, and in fact, repressed by the powers that be. Even at that early period there were powerful interests working to keep these discoveries secret. The last thing the oil companies or the energy cartels want is free energy for everybody! That would mean their own demise.

Through scalar electromagnetics it will become possible to engineer physical reality directly, even to engineering at the molecular level, creating new "impossible" molecules and even transmuting elements. It will be possible to make radioactive waste non-radioactive. This new science will unlock the secret of gravity, and thereby the secret of anti-gravity allowing us to manufacture our own "UFOs." It will allow the cure of cancer and Aids. I guess if I had to name the four most important implications of the discovery of longitudinal waves it would be these:

1. The solutions to the energy crisis and the "oil problem" are in hand. These oil-wars are completely unnecessary. There is endless energy available freely from the domain of time.

2. Unbelievably powerful weapons are not only possible, but are already operating in several nations. The many powers of these weapons are unprecedented and mind-boggling.

3. The cure of diseases such as cancer and AIDS, in fact nearly any disease, has become possible within a few years of sufficient funding. Everyone can be made healthy and stay healthy.

4. Mind control on a mass scale has now become possible, and the machines to do it are already in place in certain nations. It has become possible to mentally enslave whole populations with the twist of a few dials.

Yes, that's right, the new waves are in fact the kind of waves the mind itself uses in its operations, and so there will be implications in the realm of both mind control and education. Thoughts/feelings can be induced, even from afar, in target subjects, portending possible terrible uses in mentally enslaving people, or putting them into a happy trance, or making them apathetic, or even inducing sheer fear and panic. On the other hand, it will be possible to learn about any subject by merely "downloading" the knowledge directly into the human mind. A person might become a PhD. in a matter of months rather than years.

Other more offbeat possibilities would be "drug machines" which could induce a variety of euphoric happy states, a wide spectrum of consciousness altering effects which allow all drug users to get off their drugs. Why use drugs when those effects are available electromagnetically at the turn of a dial? Once the mind grasps the fundamental discovery it just runs wild thinking of the vast spectrum of things which will become possible.

But for mankind at its present juncture two aspects of longitudinal waves are supremely important: the matter of free-energy, and the terrible scalar weapons. If the government continues to suppress and hold back the secret of free energy from the vacuum, then the world continues to be locked into this direction we are going of the oil wars, as various powers seek to secure all the remaining oil supplies for themselves. In reality, the cost of the current war could easily have covered the research and development which would make the whole planet free of the need to burn oil! We are now spending 4 billion dollars a month in Iraq to take over their oil to the benefit of corporate America. The real benefit to the people of America would be to have a Manhattan-style project to develop scalar energy on a large scale and begin the changeover to a new oil-free era of world prosperity. But this will wipe out the oil companies and they are demons that will not go down without a fight.

The Disclosure Project of Dr. Stephen Greer seeks, in part, to force disclosure by the secret agencies of the government to disclose this technology, which it holds highly classified, for the betterment of humanity. Witnesses, many of them military or ex-military, have given testimony that such technology exists and is being held secret from the people. This may classify as one of the greatest crimes against humanity, and history may very well judge it to be so. While keeping this technology of free-energy secret, they are secretly developing it into the worst weapons the world has ever known. It is epic evil worthy of Dr. Faust.

The secret testing of these scalar devices may also explain a lot of mysteries going on, like the crop circles and the glowing orbs of lights in the sky increasing seen throughout the world. According to Tom Bearden, it was a scalar accident that caused the Chernobyl disaster. He also points out quite a number of sightings by trained pilots of "cold explosions" at sea which send up huge mushroom clouds of water vapor miles into the sky, or other sighting of giant glowing hemispheric electromagnetic domes, also called "Tesla shields."

So once again, as at the dawn of the nuclear age, mankind is faced with a discovery that could lead to its own destruction, in the hands of evil or stupid men, or could lead to an incredibly better world, free of fossil fuels, filled with healthy people, and with plenty enough of everything to go around.

Reference Article - Scalar Energy - A Completely New World Is Possible, By Bill Morgan wmorgan@nycap.rr.com. Posted 7-22-3. Extracted from the Internet January 6,2019 @ https://rense.com/general39/scalarenergy.htm[xxiii]

Turning physics on its ear - Thane Heins

Thane Heins is nervous and hopeful. It's Jan. 24, a Thursday afternoon, and in four days the Ottawa-area native will travel to Boston where he'll demonstrate an invention that appears – though he doesn't dare say it – to operate as a perpetual motion machine.

The audience, esteemed Massachusetts Institute of Technology professor Markus Zahn, could either deflate Heins' heretical claims or add momentum to a 20-year obsession that has broken up his marriage and lost him custody of his two young daughters.

Zahn is a leading expert on electromagnetic and electronic systems. In a rare move for any reputable academic, he has agreed to give Heins' creation an open-minded look rather than greet it with outright dismissal.

It's a pivotal moment. The invention, at its very least, could moderately improve the efficiency of induction motors, used in everything from electric cars to ceiling fans. At best it means a way of tapping the mysterious powers of electromagnetic fields to produce more work out of less effort, seemingly creating electricity from nothing.

Such an unbelievable invention would challenge the laws of physics, a no-no in the rigid world of serious science. Imagine a battery system in an all-electric car that can be recharged almost exclusively by braking and accelerating, or what Heins calls "regenerative acceleration."

No charging from the grid. No assistance from gasoline. No cost of fueling up. No way say the skeptics.

"It sounds too good to be true," concedes Heins, who formed a company in 2005 called Potential Difference Inc. to develop and market his invention. "We get dismissed pretty quickly sometimes."

It's for this reason the 46-year-old inventor has learned to walk on thin ice when dealing with academics and engineers, who he must win over to be taken seriously. Credibility, after all, can't be invented. It must be earned. "I have to be humble. If you say the wrong thing at the wrong time, you can lose support."

The creation in question is a new kind of generator called the Perepiteia (read related story "Holy crap, this is really scary"), which in Greek theatre means an action that has the opposite effect of what its doer intended. Heins torques up the definition to mean "a sudden reversal of fortune that's a windfall for humanity."

Deep down, Heins has high hopes. But he also realizes that merely using those controversial words - "perpetual motion" - usually brands a person as batty. In 2006, an Irish company called Steorn placed an advertisement in The Economist calling on all the world's scientists to validate its magnet-based "free energy" technology.

Steorn was met with intense skepticism and accused of being a scam or hoax. Seventeen months later the company has failed, despite worldwide attention, to prove anything under scrutiny. Well-educated people, from Leonardo da Vinci to Harvard-trained engineer Bruce De Palma (older brother of film director Brian De Palma), have made similar claims of perpetual motion only to be slammed down by the mainstream scientific community.

Heins has an even greater uphill battle. He isn't an engineer. He doesn't have a graduate degree in physics. He never even finished his electronics program at Heritage College in Gatineau, Quebec. "I have mild dyslexia and don't do well in math, so I didn't do very well in school," he says.

What he does have is a chef's diploma, and spent time as chef at the Canadian Museum of Civilization before launching his own restaurant in Renfrew called the Old Town Hall Tea Room. He has also had political ambitions. In 1999 he ran unsuccessfully as a candidate for the Green Party of Ontario, deciding a year later to run as an independent in the federal election.

Today, Heins is focused on showing his invention to anybody willing to see it, in hopes that somebody smarter than him will give it credibility. His long-time friend, Kim Cunningham, manager of communications and government relations at the Ottawa Centre for Research and Innovation (OCRI) is working part-time with Potential Difference to help get the message out.

Together, they have demonstrated the Perepiteia to a number of labs and universities across North America, including the University of Virginia, Michigan State University, the University of Toronto and Queen's University.

"It's generally always the same reaction," says Heins. "There's a bit of a scramble on the part of the observer to put what they're seeing into some sort of context with what they know. They can't explain it. They don't know what it is."

He'd be happy if somebody did, even if the news was bad. His wife has kicked him out. He doesn't earn an income. He can't pay child support. The certainty would be welcome. "I've tried to quit many times, and thought if I could just be a normal guy I would have a normal life ... But I had this idea and I believe it works."

Others want to believe – or at least help out. Cunningham, whose brother is general manager at Angus Glen Golf Club, introduced Heins to the club's president, Kevin Thistle. For two years Thistle has acted as angel investor, providing start-up capital needed to incorporate Potential Difference, file patents and continue research.

Cunningham's boss, OCRI president Jeffrey Dale, helped open doors at the University of Ottawa and make introductions to its dean of engineering. As a result, Heins teamed up last fall with Riadh Habash, a professor at the university's school of information technology and engineering.

"Dr. Habash has essentially rolled out the red carpet," says Heins, explaining that he now has access to a university lab and all the equipment he needs to test and simulate his generator.

In an interview with the Toronto Star, Habash was cautious but matter-of-fact with what he's seen so far. "It accelerates, but when it comes to an explanation, there is no backing theory for it. That's why we're consulting MIT. But at this time we can't support any claim."

In the meantime, Heins has been on a letter-writing campaign to raise money for his mission. He's written former U.S. vice-president Al Gore, Virgin Group founder and billionaire Richard Branson and John Doerr at venture capital powerhouse Kleiner Perkins Caufield & Byers. He's also tried to contact entrepreneur Elon Musk, chairman of electric car upstart Tesla Motors, and the "ReCharge IT" project run by Google's philanthropic arm.

So far no bites, though there have been nibbles. Heins has had discussions with a well-known investor in Oregon, known to many as the "godfather of start-ups," who is apparently flirting with the idea of investing in Potential Difference. "We got the impression ... he's not necessarily interested in making a ton of money, he just wants to see us succeed."

Just before the big day at MIT, the Star spoke with professor Markus Zahn about what he expected to observe.

"It's hard for me to give an opinion," said Zahn, who admitted he was excited to see the demonstration. "I don't believe it will violate the laws of physics. You're not going to get more energy out than you put in."

He said it's easy for people to set up their tests wrong and misinterpret what they see. "You've got to look closely."

It's now Jan. 28 – D Day. Heins has modified his test so the effects observed are difficult to deny. He holds a permanent magnet a few centimeters away from the driveshaft of an electric motor, and the magnetic field it creates causes the motor to accelerate. It went well.

Contacted by phone a few hours after the test, Zahn is genuinely stumped – and surprised. He said the magnet shouldn't cause acceleration. "It's an unusual phenomenon I wouldn't have predicted in advance. But I saw it. It's real. Now I'm just trying to figure it out."

There's no talk of perpetual motion. No whisper of broken scientific laws or free energy. Zahn would never go there – at least not yet. But he does see the potential for making electric motors more efficient, and this itself is no small feat.

"To my mind this is unexpected and new, and it's worth exploring all the possible advantages once you're convinced it's a real effect," he added. "There are an infinite number of induction machines in people's homes and everywhere around the world. If you could make them more efficient, cumulatively, it could make a big difference."

Driving home – he can't afford to fly – Heins is exhausted but encouraged. He says Zahn will, and must, evaluate what he saw on his own terms and time. What's preventing the engineer from grasping it right away, he says, is his education, his scientific training.

Step by step, Heins is making progress, but where it will all lead remains uncertain.

Reference[xxiv] Article - Turning physics on its ear. Written by Tyler Hamilton, Staff Reporter for Star Business Journal. Posted Mon., Feb. 4, 2008. Extracted from the Internet January 13, 2019 @ https://www.thestar.com/business/2008/02/04/turning_physics_on_its_ear.html

Section Five

Nikola Tesla

Forgotten American Scientist - Nikola Tesla

John Wagner, school teacher from Ann Arbor, Michigan, mobilized his third grade classes to raise thousands of dollars in funds to create commemorative bronze busts of Tesla to be placed at major American Universities. Many busts are already placed at many Universities.

Mr. Wagner initiated the letter writing campaign to restore Nikola Tesla to his rightful place in historical records. Here is John Wagner's story in his own words talking about the Tesla project;

This is our bronze bust (valued at $6,000) -- duplicates of which we donated to nineteen major universities. R. Farrington Sharp, sculptor

OUR PURPOSE IS -- to remind the academic community of Tesla's monumental contributions in electrical science.

THE FOLLOWING NINETEEN UNIVERSITIES DISPLAY OUR BUST OF TESLA -- THUS ACKNOWLEDGING HIS EMINENCE IN ELECTRICAL SCIENCE.

1. Harvard (Physics Dept.)

2. Yale (EE Dept.)

3. Princeton (EE Dept.)

4. MIT (EE Dept.)

5. Cal Tech (Physics Dept.)

6. U. Michigan (EE Dept.)

7. U. Wisconsin (Physics Dept.)

8. U. Maryland (Physics Dept.)

9. U. Illinois (EE. Dept.)

10. Purdue U. (EE. Dept.)

11. U. Pennsylvania (Physics Dept.)

12. Georgia Tech (EE Dept.)

13. Cornell U. (Physics Dept.)

14. Michigan State U. (Physics Dept.)

15. Johns Hopkins U. (Physics Dept.)

16. Penn State U. (Physics Dept.)

17. Duke U. (EE Dept.)

18. Carnegie Mellon U. (EE Robotics Dept.)

19. Columbia U. (EE Dept.)

THESE ARE THE WORLD'S TWO MOST IMPORTANT TECHNOLOGIES:

It was Tesla (not Edison) who gave the world **electromagnetic power (AC)**.

It was Tesla (not Marconi) who gave the world **electromagnetic communication (Radio)**.

Yet Tesla's name is:

Omitted in school textbooks

Omitted in technical journals

Unknown even to some engineers

and

The importance of his inventions minimized -- the Smthsonian even omits his name from *The Smithsonian Book of Invention* and *The Smithsonian Visual Timeline of Inventions*

also

"TIME" magazine credits Thomas Edison for changing America. It wasn't his lightbulb that changed America -- it was Tesla's breakthrough discoveries in AC power transmission and radio communication that caused the meteoric explosion of progress commencing in 1896 and continuing today at an ever expanding rate.

OUR FIRST EFFORT

1. •Verify Tesla's eminence in electrical science to the American public by donating his bust to major universities, funded by selling promotional T-shirts and receiving many donations for 21 years.

1. •We still need to achieve a 'critical mass' of Tesla's bust donations to major universities.

OUR NEW EFFORT

1.•Introduce Tesla to younger students with our two biographical books of his childhood written for children.

Nikko & Macak, the Electric Cat, (Copyright 2010 by John W. Wagner), ISBN 978-0-615-36681-4, LCCN 2010905209

A 23-minute CD narration is included for children ages 5 - 7. Children age 8 - 9 can read the story themselves, or listen to the CD.

Introduction

This is a true story of how the American inventor, Nikola Tesla, deliberately trained himself at a pre-school age into becoming a virtual human invention machine.

It describes his early childhood experiences that made him so different from his playmates. It tells how he first discovered electricity by stroking his pet cat's fur, causing a violent and memorable reaction. It was this experience that led him to become an inventor.

It tells the experiences he had with his two doting aunts and how he was able to gain control of his will in dealing with them. It tells also of his experiences with farm animals. Tesla died in 1943, alone and a virtual pauper, but surrounded by an advanced technological world his scientific discoveries in electrical science had made possible: AC power and radio.

Nikko, Superboy Inventor, (Copyright 2010, by John W. Wagner), ISBN 978-0-615-39715-3, LCCN 2010912313. Recommended for children ages 9 - 14. Older children may prefer reading a Tesla biography written for adults, such as *Tesla, Man Out of Time* or *Prodigal Genius, the Life of Nikola Tesla*.

The reader will see that Nikko demonstrated unusual talents and a kindhearted, benevolent nature. The story shows how his unique experiences influenced him in his quest to benefit mankind by discovering nature's hidden energy secrets. It tells how he interacted with his family and playmates; how, at a preschool age, he made a waterwheel to test his first theory that the energy of moving water could be changed into a rotary motion; how he almost destroyed his town with a snowball; how he made an engine powered by insects; how he captured and raised an eaglet so he might learn the secret of flight, then made wings to test his plan. It relates the struggle he had with his priest father who wanted him to enter the priesthood instead of pursuing a rigorous education in science; and it tells how that struggle drove him to the brink of death.

This is a compelling story of a boy possessed with an insatiable curiosity and desire to solve a problem the world believed to be impossible – tantamount to perpetual motion. In the end, he solved the conundrum of alternating current, and put into motion the Second Industrial Revolution commencing in 1896. AC power made its debut at the Niagara Falls power station that year and the world was changed forever. AC power generating plants soon began springing up throughout the world, bringing technological advancements in all areas to a civilization that would have previously believed them to be impossible. Today, the world has all but forgotten Nikola Tesla, who from childhood wanted only to make the world the beneficiary of his superior intellect.

Readers, we are not talking about an obscure inventor lost in history, nor are we talking about a mad scientist who performed magic. Nevertheless, in the nineteenth century when he made his early discoveries, they may have seemed like magic.

We are talking about a man whose major scientific gifts to the world (of AC power transmission and Radio), more than anything else, caused a Second Industrial Revolution. How else can one explain the meteoric advance of science from the horse an buggy days of 1895, when the Niagara Power Project adopted his polyphase AC power system, followed by his invention of Radio...then a mere 74 years later man walked on the moon? It just had to be Tesla who 'invented tomorrow.'

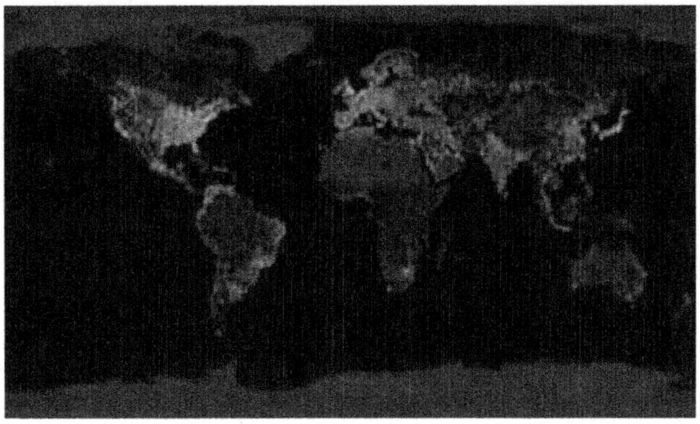

At present we have about 900 GW of generating capacity in the USA, far more than any other country. This map above of electric generating stations clearly shows Tesla's influence on the world.

Trust me, it was not Mr. Edison's light bulb that caused this revolution, as the Smithsonian Institution wants everyone to believe; it was Tesla's AC motor, his polyphase AC power transmission of electricity, and his invention of Radio.

His AC motor brought unprecedented mechanical power to industry; his AC power distribution system made electrical energy available everywhere; and his radio communication system allowed us to communicate ideas with each other.

Is it not an absurdity that this great genius is virtually expunged from history in our country? It is more than that; it is a national disgrace!

HIS NAME MARKS AN EPOCH IN ELECTRICAL SCIENCE

In 1882 he made the discovery that changed the world...harnessing the awesome power of Alternating Current.

Each of these high tension power line towers throughout the world is a monument to Tesla; his genius created them.

In 1888 he obtained U.S. patents covering an entire system of polyphase AC that remains unchanged in principle today.

He promptly sold his patents to George Westinghouse, an acquisition that made the Westinghouse Company the giant it is today.

Westinghouse and Tesla were consummate friends, but after Westinghouse died in 1913, the company forgot about its chief benefactor and Tesla fell victim to hard times.

He died January 7, 1943, alone, poor, and all but forgotten, in his only home, a New York hotel room. This web site is his story...and ours.

After Tesla's discovery of the rotating magnetic field principle, making virtually unlimited electric power available to the world, he moved on to finish the work of Maxwell and Hertz by inventing first his Tesla coil, and then the four-tuned circuit, making radio a viable means of communication.

UPHELD BY THE SUPREME COURT

Tesla's four-tuned circuits (two on the receiving side and two on the transmitting side, secured by U.S. patents #645,576 and #649,621) were the basis of the U.S. Supreme Court decision (Case #369 decided June 21, 1943) to overturn Marconi's basic patent on the invention of radio.

Marconi merely demonstrated Tesla's invention, but the gullible media and the greedy industry that followed continue to perpetuate a myth that Marconi invented radio. Whom do you believe has more credibility...industries promoting their own businesses, or the U.S. Supreme Court?

Marconi's two-tuned circuit system was the same as that advanced by Heinrich Hertz and was no more a viable system of radio than that advanced by Mahlon Loomis in 1872...long before Hertz or Tesla. In one of my pages I tell the complete story in legal and technical terms. Any unbiased reader should understand that it was Maxwell, Hertz, and Tesla who were the actual pioneers of radio. Marconi only helped develop radio after it was invented. Mr. Edison and Mr. Marconi's popularity in history is just another example of historians promoting entrepreneurs and technologists over the actual discoverers.

Today, industries flourish and the world surges from the power his fertile mind created...and radios blare with news and music, their transmissions made possible by his giant intellect...all telling us that TESLA WAS HERE.

Michael Faraday rocked the scientific world in 1831,with his discovery that magnetism can produce electricity...if it is accompanied by motion. Faraday discovered the principle, but not how to make it power the world; Tesla alone accomplished this singular feat.

Tesla is the greatest inventor the world has ever forgotten. He is also the greatest inventor the Smithsonian has ever 'swept under the carpet.'

He holds over forty U.S. patents (circa 1888) covering our entire system of Polyphase Alternating Current (AC). These patents are so novel that nobody could ever challenge them in the courts.

The Direct Current (DC) system Edison used in his much touted Pearl Street generating station was invented by others before his time; he merely copied the work of others to promote his business enterprise...and the Smithsonian wants you to believe he was America's 'King of Electricity.' There is simply no evidence to support this claim.

Lest you jump to the wrong conclusion, we are not criticizing Mr. Edison whose Menlo Park Laboratory workers, under his direction, made many practical inventions; we are criticizing only the groups promoting Mr. Edison's name in the electric power industry; he actually fought against the adoption of Tesla's AC system.

OUR QUEST

After we learned the heart wrenching story of Tesla's life and realized what a great American he was, we wanted to tell the world about him. We wrote many letters to important people asking for their support.

Then a former student approached me one day. She said she had persuaded her father, an accomplished sculptor, to create a bust of Tesla for our class, but we had to pay for the materials.

The sculpture is bronze, mounted on premium grade solid granite imported from India. Its appraised value is $6,000. The plaque reads as follows:

WE WROTE MANY LETTERS

Cursive handwriting is normally introduced in the Third Grade. Now our classwork had a purpose...WRITING LETTERS to raise money for our Tesla bust.

This is a photograph of the letters we wrote to more than one hundred American corporation presidents and CEO's. Many responded with generous checks to support our campaign. We even received a check from the president of Sony Corporation in Japan.

We offered our bust to the Smithsonian, but Dr. Bernard S. Finn, (then Curator of the Division of Electricity and Modern Physics) refused us, claiming he had no use for it.

We could not understand why the Smithsonian would have no use for a $6,000 bust of such a great American and eminent scientist who stands alongside such other great scientists as Volta, Watt, Ampere, Ohm. and only ten others in the world.

HONORED BY THE WORLD

Names for units of electrical measurement are derived by using the names of scientists who made the greatest contributions in electrical science, forming perhaps the most elite group in the world.

In 1956, Tesla's peers at the Electrotechnical Conference in Munich acknowledged his monumental contributions to science by designating his name to represent a unit of magnetic measurement. Thus, the "TESLA" became the Unit of Magnetic Flux Density in the MKS system.

Throughout the entire history of electrical science only fifteen men worldwide have received this honor. Tesla is one of these great men. (Edison and Marconi are NOT!)

In addition, Tesla received fifteen honorary degrees from famous universities worldwide, including Yale and Columbia in the United States. He also received fourteen Awards of Merit from other eminent groups.

Dr. David L. Goodstein, Vice Provost and Professor of Physics at California Institute of Technology, calls Tesla one of the "Saints of Science" and equates him to Leonardo Da Vinci.

TESLA ERASED AT THE SMITHSONIAN

THE BRAIN WASHERS

I urge you to continue reading our web page for a richly interesting story of how our country's premier museum, the Smithsonian Institution, deceives the public by writing biased history...and their curator is doing this in consort with the History Committee of IEEE (Institute of Electrical and Electronic Engineers) at Rutgers University.

NOT ALL GUILTY

The majority of IEEE members are oblivious of the Smithsonian and the IEEE History Committee's biased agenda. Nevertheless, everyone who does not speak out against their wrongful depiction of electrical history deserves having their children spoon-fed the popular corrupted version.

A TRIP TO THE SMITHSONIAN

When I visited the Smithsonian to learn why Dr. Finn had no use for our Tesla bust, the reason became indelibly clear.

Please remember, at this time I was still naive about the Smithsonian's bias against Tesla; that is, until I saw......

Yes, bust of Edison!

Next to Edison's bust I saw Tesla's invention. This is his invention that revolutionized the world.

Here is a photograph of Tesla's rotating magnetic field device I saw, giving us polyphase AC and the AC motor...

Tesla's U.S. patent number was on his invention, but I could not find any recognition of Tesla. When I asked Dr. Finn why he had placed Edison's bust on display next to Tesla's invention, he said, "The sculptor was a phrenologist and wanted to examine the bumps on Edison's head; this makes our display authentic."

The entire electrical display at the Smithsonian (including their web site) focuses on Edison's brief business enterprise that failed. The Smithsonian is not telling a story of history and invention; they're perpetuating a myth.

Edison used Direct Current (DC), a technology invented and developed by others (before his time), as a means of powering his incandescent lamp. Zealous industry and the gullible media have exaggerated this story so much that now everyone believes Edison is the father of our system of electrical power.

Please continue reading to learn how you can help correct history. Trust me...this is the ONLY way. We must reintroduce Tesla to the academic community through students. We started with our country's leading universities and hope to continue down to our elementary schools.

Are you still not convinced that our complaints are legitimate? Please consider the following information concerning the Wright Brothers...not willingly remembered by historians.

SMITHSONIAN SNUBS WRIGHT BROTHERS

When you visit the National Air and Space Museum at the Smithsonian you see this famous Wright Flyer that made man's first successful flight December 17, 1903. What Smithsonian officials do NOT tell you is that they snubbed the Wright brothers for 45 years, refusing to acknowledge their great accomplishment and install this famous plane in the museum. They refused because their own Secretary of the Smithsonian, Samuel P. Langley, built an airplane shortly before the Wright brothers...but it could NOT fly!

Forty-five years is a long time for the Smithsonian to deny the truth. Wilbur died Spring 1912, weakened by his nine-year dispute with the Smithsonian. Orville finally gave up the fight in 1928 and sent his famous plane to the Museum of London as a gesture of contempt for the Smithsonian.

American public pressure increased in the years that followed. Many people wondered why our famous Wright Flyer was in London instead of here in

America. Orville died January 1948. Later that year the Smithsonian finally agreed to bring the plane back from London to be formally installed December 17, 1948. Unfortunately, neither of the Wright brothers lived long enough to know that their own country officially acknowledged their great accomplishment.

The Smithsonian is cheating the public from learning Tesla's history in much the same way as they did with the Wright Brothers, except that Tesla's snubbing has lasted more than a century.

HERE IS ANOTHER EXAMPLE

Under pressure from some members of Congress to pay tribute to Tesla, our government issued a block of stamps in 1984, casting him alongside other scientists whose work was of much less importance than Tesla's discovery of the rotating magnetic field idea.

They credit Tesla only with inventing the AC motor, and fail to specify that the AC motor in reverse is a polyphase AC generator, the rock foundation of virtually all power generated in the world. Further, none of the other three men have units of electrical/magnetic measurements named in their honor. In short, they are not even close to being in Tesla's league.

The government also fails to recognize the US Supreme Court's landmark decision of June 21, 1943, Case # 369, overturning Marconi's basic patent for the invention of radio, thus proclaiming Tesla as the real inventor of radio.

WE HOPE YOU ARE INCENSED ENOUGH TO WANT TO HELP US?

Education is the only way to combat the Smithsonian's wrongful depiction of electrical history, but we cannot hope to match the millions of dollars industry and the Edison Institute spend promoting Edison's name. Nevertheless, I believe we can make a significant impact on many of our country's future physics and engineering students by donating busts of Tesla to many of our major universities.

People universally recognize that a sculpture is an acknowledgment of one's greatness. Inasmuch as the academic community essentially forgot Tesla for a century, we believe a good first step is to reintroduce him to the students and faculty of our major universities. Our second step is to introduce him to grade school students with our books for children.

OMITTED IN SMITHSONIAN BOOK OF INVENTION

The Smithsonian Book of Invention is an extra-large hardcover book almost an inch thick. Many inventors and their inventions are shown and their impact on civilization discussed...such people as Einstein, Bell, Goodyear, Edison, and many others, but Tesla and his epic-causing discoveries are omitted.

Nevertheless, strange as it may seem, the book included Archie Bunker and Colonel Sanders.

The Smithsonian's Visual Timeline of Inventions Book cites Rubik's cube, the electric toothbrush, and the pop-up toaster, but fails to list the AC motor. Tesla is not even listed in the Index. Further, they credit the invention of radio to Guglielmo Marconi---"1895...After reading the scientific writings of Heinrich Hertz, 20 year old Italian Guglielmo Marconi invented radio communication."

The Smithsonian ignores Thomas Commerford Martin's biography of Tesla, published in 1894, describing Tesla's demonstration of radio transmission in 1893. The Smithsonian also ignores (as previously mentioned) the U.S. Supreme Court's decision upholding Tesla's patents for the invention of radio.

Is it any wonder many Americans have lost faith in the Smithsonian? Their leaders are not only writing biased electrical history, they are systematically dismantling our country's historical heritage and replacing it with trash to promote a self-serving agenda...resulting in history revisionism.

CREDIT GIVEN TO EDISON FOR TESLA'S INVENTION

Dr. Bernard S. Finn was curator and first author of this Smithsonian publication. In his section entitled "The Beginning of the Electrical Age," he names 43 contributors to the science of electricity. Mr. Edison's name is cited many times along with his photographs, but Nikola Tesla's name is omitted.

Equally outrageous is the Niagara Falls power station picture of Tesla's AC generators on the last page...and Dr. Finn's concluding remark: "When the Niagara Falls power station began operating in 1895, it signaled the final major act in the revolutionary drama that began in Menlo Park in the fall of 1879."

BRAINWASHED

By this time the totally brainwashed reader is led to believe that our electrical world started with Mr. Edison at Menlo Park when he invented the light bulb in 1879; then he finished electrifying America in 1895 by creating the Niagara Falls power station. Yet it was Tesla's nine basic U.S. patents that were used in that power plant's creation. Edison had no role in that project whatsoever.

PHOTO OF PLAQUE ON ORIGINAL GENERATOR

Edison actually fought the adoption of AC bitterly by waging his infamous "War of the Currents," culminating in his creation of...

THE FIRST ELECTRIC CHAIR (1890)

Yes, it was Thomas Edison who invented the electric chair to frighten people away from the use of Tesla's AC system of electricity.

EDISON ELECTROCUTED ANIMALS FIRST

This drawing (circa 1889) is that of a horse being electrocuted in Edison's Menlo Park laboratory. Edison was also known for paying children 25 cents for each stray dog they could bring him. Then he held press conferences and electrocuted the dogs at public gatherings in a desperate attempt to frighten people away from using AC. He claimed DC could not kill...but in truth, DC could also kill.

Edison felt it was necessary to experiment killing animals before he could guarantee his electric chair would kill efficiently.

THE IRONY OF IT ALL

Is it not classic irony that today Americans hold Edison in such high esteem, many even paying their electric bills to companies bearing his name, while Tesla, our actual benefactor, is essentially erased from history...in our schools, in our technical journals, and even in our marvelous Smithsonian Institution? Something has gone awry and we need to correct this sorrowful situation.

ANALYSIS OF NATIONAL MUSEUM OF AMERICAN HISTORY HALL OF ELECTRICITY HOME PAGE

WHAT IS THE PROBLEM?

As in many institutions, the Smithsonian is not without its problems -- and money is often at the root. Approximately 83% of its operating funds comes from the federal government; the remaining 17% comes from donations and what can be earned from grants from American industries. When an industry gives money to the Smithsonian, the curators feel obliged to favor the history of that industry as much as possible in their displays or publications.

CLASSIC EXAMPLES

Two examples of this assault on history are: (1) The Edison Institute (2) The Orkin Company (you know, the bug killers). In the case of the Edison Institute, they routinely give money to have Smithsonian curators write a story and create a display featuring Edison and his invention of the incandescent lamp. In depicting the history of Edison's incandescent lamp, the curator loses perspective of historical events and omits critical parts in order to show favor to The Edison Institute.

LIGHT IS NOT AS IMPORTANT AS ELECTRIC POWER AND MECHANICAL ENERGY!

The curator hails the invention of the light bulb as being the cause of the Second Industrial Revolution. It was NOT! Nikola Tesla's invention of the rotating magnetic field, (giving the world AC power and the AC motor) was clearly what American industry needed to bring about that revolution; it provided the necessary muscle to do the work. This is not to say that light was not a welcome advancement, but its importance pales in comparison to AC power and AC motor.

THE SMITHSONIAN MADE STORY CHANGES

Earlier I criticized the Smithsonian's flagrant omission of Tesla's work and their deceptive rhetoric. I based my criticism on their web site entitled "Lighting a Revolution." Recently I discovered the Smithsonian's revised web site; it is now called "Lighting a Revolution II." Tesla is now mentioned, but even the casual observer will notice that Edison's name dominates the Smithsonian's many articles.

THE TALL TALE CONTINUES

The curator traces the history of Edison's business enterprise (of building direct current [DC] power stations) to power his incandescent lamp. In doing so he creates the false illusion for a naive citizenry that Edison was the inventor of electricity...when, in fact, he was merely a businessman using a technology others had already developed. Direct current technology was dead from the beginning because DC could be sent only over short distances.

Readers should take a moment to examine the National Museum of American History's "Hall of Electricity" page.

ANALYSIS OF THEIR OWN WORDS:

The following are direct quotations from the "Hall of Electricity" web page: "The displays help illuminate how Thomas Edison's lightbulb and other inventions began to transform our world."

Notice how the curator gives emphasis to Edison's lightbulb, but omits naming the "other inventions." He even uses the word "illuminate" to describe what the displays will do for the observer, further leading the reader to believe that Edison and his lightbulb "transformed the world."

OMISSIONS TELL A STORY TOO

"The rest of the hall is devoted to a separate exhibition, Lighting a Revolution. At the end of the19th century, a new way of transmitting power came to the fore -- electricity. In 1876, at the Philadelphia Centennial Exhibition, a few arc lights were shown. They were very bright, suitable only for large open spaces. Three years later, Thomas Edison announced his invention of the incandescent light bulb, and on New Year's eve in 1879 drew a crowd of 3,000 visitors to his Menlo Park, New Jersey complex to see the buildings and grounds aglow in the softer light of his creation. Lighting a Revolution explains in detail the complex circumstances behind Edison's revolution."

Notice how the curator emphasizes light. Also, he talks of "a new way of transmitting power at the end of the 19th century." No doubt, the naive or casual reader automatically thinks the curator is referring to AC, and in doing so they identify Edison with AC because many power companies have Edison as their namesake.

THE GUIDING LIGHT

"Edison and his colleagues produced not merely a light bulb but an electrical system, and the next section tells the story of his early power plants, including the first central at Pearl Street in New York City, which began producing power on September 4, 1882."

Again, the curator carefully steers the reader to believe that Edison is responsible for every advancement made in electricity. He also carefully steers the reader to believe that Edison's lamp created the revolution.

EDISON'S FAILURE

"Edison's success drew competitors, some of whom improved on his methods, especially through the introduction of alternating current, which ultimately supplanted his direct-current system. The next section of the exhibition shows the lamps, meters, generators, and other devices of some of these competitors and explains the reasons why alternating current proved superior to direct current."

The curator calls Edison's enterprise a success...when, in fact, it was NOT. Tesla tried to give (yes, give) his AC system to Edison because he realized direct current technology was flawed, but Edison stubbornly refused to acknowledge the superiority of Tesla's AC system. Notice how the curator carefully avoids using Tesla's name and refers to him only as a 'competitor'. It was Tesla who held over 40 U.S. patents on a complete AC system that proved to be the breakthrough discovery the world thought was impossible. Note also how the curator identifies direct-current as "his" (Edison's) system, when in fact direct-current technology was already fully established seven years before Edison hired electrical engineers to build his much touted Pearl Street Generating Station using this established DC technology.

THIS IS TOO MUCH!

"An Edison electric motor of about 1890. In 1895 a large remote generating station began producing electricity at Niagara Falls. Less than two decades after Edison's invention, electricity thus was accepted as a principal means of power transmission. A revolution had taken place. Cheap electric power made new industrial processes possible, such as the economical production of aluminum. Eventually this power reached the city and the home, where its influence is made clear in a case filled with early 20th-century appliances such as fans, coffee pots, and vacuum cleaners."

First, the curator shows a picture of an "Edison" electric motor of about 1890, setting the stage for continued glorification of Edison. Tesla created a worldwide revolution in electrical science two years earlier (1888) when he introduced his new AC motor, yet the curator persists in talking about an 1890 old technology motor of which Edison had made no scientific contribution. In the next sentence, after displaying an "Edison" motor, which he did not invent the curator tells of the Niagara Falls power generating station. The curator fails to explain that the Niagara plant was designed to generate AC power using Tesla's system--Tesla's patents had to be licensed to build it. Then he carefully avoids identifying Tesla as the creator of this project. Instead, the curator states, "less than two decades after Edison's invention, electricity thus was accepted as a principal means of power transmission." The curator, by using deceptive semantics and omissions, leads the reader to believe that two decades after Edison's invention of the "light bulb," Edison introduced electrical power to the world. Then he reinforces his statement by stating that "a revolution had taken place." (implying that Edison had caused that revolution.) The curator is correct in stating that a revolution had taken place, but he failed to credit the person who caused that revolution...Nikola Tesla. In the curator's next statement he tells how "cheap electric power made new industrial processes possible, such as the economical production of aluminum." The above statement is a particularly flagrant deception. The production of aluminum requires an enormous amount of electricity that only AC can provide. The curator, by omitting Tesla's name as the creator of polyphase AC, guides the reader to believe that Edison was responsible for the electricity needed to germinate the aluminum industry. (Of course, the aluminum industry is what spawned the aircraft industry...all made possible by Tesla's AC power.) The curator continues to describe how electricity (presumably from Edison) brought numerous consumer items to market...citing fans and coffee pots. He carefully neglects to mention that most all small appliances use Tesla's AC motor. This curator, in his zeal to highlight Edison, creates the false illusion that Edison was the inventor of our entire electrical world.

WHAT EDISON ACTUALLY DID

Edison's greatest creation was his invention factory at Menlo Park, NJ. From that factory came many wonderful and practical inventions, but electrical power was NOT one of them. The curator is dead wrong in claiming there was creativity in Edison's power enterprise. There was creativity from the Menlo Park group for the invention of the incandescent lamp, the phonograph, the moving picture machine, and other popular devices, but not electric power! Edison merely expanded upon a known technology (DC) and was too stubborn (or ignorant) to recognize its limitations. Power companies that use Edison as their namesake are only perpetuating a myth; it is a pathetic joke and they need to get their history straight.

A BUG KILLER MADE IT TO THE SMITHSONIAN

In the case of the Orkin Company, they donated a half million dollars in 1992 to the National Museum of Natural History. In exchange, the Smithsonian agreed to create a permanent display honoring Otto Orkin, founder of the Orkin Company--another example of our prestigious Smithsonian catering to corporate commercialism.

WHAT IS INVENTION?

The dictionary describes invention in the classical sense...that of finding out, discovery. The Smithsonian's liberal definition of 'invention' certainly violates this definition; instead, supplanting it with entrepreneurship.

For example, if we asked the question, "who invented medicine?"

Would it be the first man who proposed theories of nature's properties, without ever knowing their possible applications?

Would it be a second man who then proved the existence of these properties--again without realizing their possible application?

Was it a third man who then built on this knowledge and made the first application of nature's properties by actually making a medicine and proving that it worked?

Or was it a fourth man who made a business from the first three men's discoveries?

While it would be grossly unfair to give the third man full credit because he used the knowledge gained by the first two, it should be unthinkable to afford any credit to the fourth man.

Such is the situation with Edison and Marconi.

They were clever businessmen who received much publicity, which usually creates false perceptions. The Smithsonian's wrongful depiction of electrical history simply enhances this perception.

CONCLUSIONS

Please do not believe that any 'one-man' invented our system of electricity, nor did any 'one-man' invent radio. The story of the rise of electricity is long and interesting, and should not be altered by biased historical interpretations of entrepreneurial experiments made by the many inventors who followed the 'Great Discoverers.' Thomas Edison is a classic example of a businessman who was successful in utilizing the talents of others to create many useful devices that became popular with people, but practical electricity came only from Nikola Tesla. Marchese Guglielmo Marconi was the first to send a message across the ocean. As a result of his business interests later, he is partly responsible for 'developing' radio...but he did NOT invent it. Men such as William Gilbert, Alessandro Volta, Hans Christian Oersted, Andre-Marie Ampere, Georg Simon Ohm, Michael Faraday, James Clerk Maxwell, Heinrich Hertz, and Nikola Tesla are the true inventors of electrical power and radio. Yet these science heroes are mentioned only in certain college courses. Therefore, their accomplishments are virtually unknown to most people. It is impossible to say which one of these great discoverers made the most significant contribution to science, but one fact is indelibly clear...Nikola Tesla made the final breakthrough in the electrical power area with his rotating magnetic field principle. He also made the final breakthrough in radio by devising his four-tuned circuit idea, which was the first viable radio system. Then he demonstrated how it would serve mankind in the future. His system is basic to all radios today. I tell about this in greater detail in the pages that follow.

THE NATIONAL INVENTORS HALL OF FAME

Another example of biased history is the NIHF web site telling our children that Charles P. Steinmetz, an engineer at the General Electric Company, is the inventor of our electrical distribution system.

Charles Proteus Steinmetz
Born April 9, 1865 - Died October 26, 1923
System of Distribution by Alternating Currents
Patent No. 533,244. Inducted 1977

This is such a preposterous absurdity that it is laughable...but I am not laughing. Instead, I am blowing the whistle on NIHF for being nothing more than puppets of industries eager to wave their own flags instead of telling factual history.

Tesla's seminal patents of 1888, on our now worldwide system of polyphase AC had to be licensed by GE to complete the Niagara Falls Power Station in 1895. This power station was the first large scale hydroelectric polyphase AC generating plant to transmit power over long distances. Mr. Steinmetz was a brilliant engineer who later worked out the laborious details of electrifying the country, but the basis of his work was founded and patented by Tesla. Yet NIHF never mentions Tesla's name in their story. It is small wonder our children's science textbooks are riddled with misconceptions and downright errors.

NIHF has inducted Tesla into their ranks as one of their notable inventors, but instead of explaining the complete story of Tesla's monumental discoveries in AC and Radio and their importance to the world, they skirt the periphery and credit him only with the invention of the AC induction motor. They fail to identify Tesla's rotating magnetic field principle--basic to the AC motor--being the rock foundation of our worldwide system of polyphase AC, without which the wheels of industry would stop and our homes would be without power. In short, the AC motor in reverse is a generator of power, but they never mention this elementary fact. To do so would be contrary to their story of Thomas Edison's role as our chief benefactor of electric power; he was NOT.

We all know what happens when the power fails...everything comes to a grinding halt. I'd say that Tesla's discovery of AC power transmission is rather important. Mr. Steinmetz merely helped implement Tesla's discovery. Nevertheless, we should not diminish his work anymore than we should ignore Tesla's work.

NIHF also credits Guglielmo Marconi with the invention of Radio and ignores the fact that the U. S. Supreme Court, in a landmark decision, overturned Marconi's basic patent on Radio because Tesla's patents had priority. Tesla's patents serve as the rock foundation of all radios today.

HOW AND WHY I BEGAN THIS QUEST

In 1983, I was browsing through the Allegheny College Library--I had traveled there to bring my daughter home for spring vacation. While there I explored their library and found a biography, PRODIGAL GENIUS: THE LIFE OF NIKOLA TESLA, by John J. O'Neill, (1944). I had known about Tesla and his famous coil from my high school days 40 years earlier, but this was my first opportunity to learn more about the man whose coil made fascinating sparks. Regrettably, his coil was the sorrowful extent of my knowledge of his work, and likely, this is as much as most people know...if indeed they recognize his name at all.

The first two pages of O'Neill's biography are powerful. After more than 27 years I cannot forget his words. In short, I deem his book of epic proportion. When I looked at the inside of the back cover, I was amazed to find the book had been taken out only eight times in 38 years. How could this be? Was I overestimating the importance of its message? I did not think so, and I still don't. The same can be said for the next Tesla biography I discovered, TESLA, MAN OUT OF TIME, by Margaret Cheney, (1981). It has a powerful introduction by Leland Anderson, our country's foremost authority on Tesla--he was Cheney's mentor. It is by far the most authoritative and well documented book available today on Tesla.

I had been teaching English Composition and my first thought was, 'how interesting it might be for my students to learn the story of Tesla's life; it just might help me breathe some fire into their writing.' The more I read, the more I was able to tell my students, and the more they became inflamed with enthusiasm...not only for writing but for a desire to wave Tesla's banner. It seemed we had discovered a secret, if not an outright flaw in history. Why was it that so few people had ever heard of Tesla? We decided to write letter stories to people and tell them about our discovery. At first, and without realizing it, Tesla had become our class hero, and my students now became excited about writing. We developed a mission...to tell as many people as possible about Tesla. I was reveling at the writing successes in my classes, so I continued this activity with succeeding classes for several years.

One school year, 1986-87, was particularly exciting. Someone brought to our attention an interesting situation that existed in Madison, Wisconsin. We learned that several years earlier someone had named streets after famous scientists, including Tesla and Marconi. The city fathers spelled "MARCONI" correctly, but somehow they had misspelled Tesla's name as "TELSA" on their street sign. We also learned that the Chairman of the History Department at the University of Wisconsin years earlier had petitioned Madison officials to correct the spelling of Tesla's name, but his pleas to City Council were rejected repeatedly until he finally abandoned his crusade. This was exactly the kind of challenge my students wanted and needed. We bombarded Madison with letters to City Council, the newspaper, residents on "TELSA" Street, the mayor, even the mayor's wife, chiding them to correct Tesla's misspelled name. We argued, 'after all, Tesla was an honored American citizen who gave much to the world.' Also, 'how would they like to see Washington's name spelled incorrectly on street signs?' We had a birthday party for Tesla at my home that summer and made posters.

Photo of our yearly birthday party for Tesla at my home

The dispute raged for months and our story made the front page of the Wisconsin State Journal. One council member was particularly distraught over the issue. He called one day and begged that we 'get off their backs.' We refused because we knew we had truth on our side. Five months later we received a letter from the City Clerk advising us they had corrected the street signs. This was confirmed later by a photograph one of my students took while traveling there.

Two of my former students sent me this photo while on vacation in Madison, WI

One Saturday morning two years later my doorbell rang. It was a former student and her father. She told me that for the past three years she had not forgotten our Tesla story. It had made such an impression on her that she had persuaded her father, Ronald Farrington Sharp, an accomplished sculptor, to render a bust of Tesla for our class... that is, if this was something I really wanted. It took about a microsecond for me to agree to their generous offer. They explained that I would have to pay for the materials, but there would be no charge for the work. I was ecstatic and many thoughts ran through my head. We could put the bust in our classroom, or perhaps donate it to some museum. When the clay model took better shape and I could see how magnificent it really was, the sculptor asked what preference I had for the final cast. We decided on bronze. We also decided on a granite base and a bronze plaque telling of Tesla's contributions to electrical science. I worked for weeks at my computer keyboard composing various inscriptions for the plaque and consulting Margaret Cheney and Leland Anderson for guidance. Finally, I decided to use the following inscription:

HIS NAME MARKS AN EPOCH

IN A SINGLE BURST OF INVENTION HE CREATED THE POLYPHASE ALTERNATING CURRENT SYSTEM OF MOTORS AND GENERATORS THAT POWERS OUR WORLD. HE GAVE US EVERY ESSENTIAL OF RADIO, AND LAID THE FOUNDATION FOR MUCH OF TODAY'S TECHNOLOGY.

There will probably be several readers who might prefer I had said something else on this plaque. Nevertheless, these were Tesla's greatest achievements.

My younger son told me about a band in California named TESLA that might be interested in contributing to our campaign, so I wrote them. In a few weeks their drummer and a guitar player responded personally with a check for $1,800. In addition, we wrote to dozens of CEO's of power companies and large corporations. Many responded with checks of $50 to $100. It seemed we were making real progress because many people in high positions were sympathetic to our message.

As we struggled, deciding where to place our bust, someone suggested the Henry Ford Museum in Dearborn. Its museum has an entire section devoted to the history of electrical science. We believed their curator would welcome a bust of this caliber for their display. When I met the head curator I showed him a large full color photograph of our bust and told him our story about how it came into being. Then, on behalf of my students, I offered the bust, fully expecting that he would accept it. After all, the bust was magnificent and had an appraised value of $6,000. I was struck dumb when he refused our offer. It was simply beyond my comprehension that such a famous museum would refuse to accept and display the bust of America's greatest inventor in electrical science. As I left the building, admittedly dazed, I noticed a huge model of the Statue of Liberty, the central feature of an Edison display. The theme was a celebration of Edison for making possible the electrical energy needed to illuminate our national monument. It was apparent AC power was celebrated, rather than Edison's incandescent lamp. As I walked out the door I looked back, and there above the entryway I saw an inscription I had apparently missed when I first entered. It read, "The Edison Institute." Now everything suddenly fit into place. Edison and Ford were close friends, and it was reasonable to expect the Ford Company would continue to celebrate Edison's name. I thought, 'there must be other museums that would not have this kind of bias,' so I visited the Chicago Museum of Science and Industry. I concluded it was nothing more than a haven for commercial displays...certainly nothing to compare with, say, the Ontario Science Centre. Canada was out of the question though because we wanted to keep our bust in this country.

Finally someone suggested we go to the top...to the most revered institution of them all, the Smithsonian. It is the one place in our country where only the greatest have their places in history preserved forever...men such as George Washington, Thomas Jefferson, and Joseph Henry. We felt their officials would understand and appreciate Tesla's merit, and wondered if they had recognized Tesla's accomplishments and already had a bust, so I decided to travel to Washington in 1988 to visit the Smithsonian's National Museum of American History. I had visions of seeing nothing but the grandest of sculptures and extensive displays. Upon arriving I discovered the electrical section was closed for renovation. When I explained to officials my purpose of coming to Washington, I was treated to a personal tour of their archives by the curator's assistant. He gave me a book describing their displays, but I was also able to see a part of the museum that was open to the public. There were numerous busts, but none were so great as to make our bust seem inferior. Therefore, I saw no reason why the museum could possibly refuse it. I returned to Ann Arbor with the book under my arm and feeling as though my trip had been a success. All that remained was to make our formal offer and our project would be ended.

When the bust was finished, one of my students wrote to offer it. In a few days we received a reply from Dr. Bernard S. Finn, then Curator of the Division of Electricity and Modern Physics. He did not want the bust! My students were devastated, and again I was struck dumb. He explained they 'almost never collect busts in their Division.' Later, another trip to the museum revealed that Dr. Finn was displaying Edison's bust next to Tesla's first AC motor/generator. Tesla's U.S. patent number appeared on the motor/generator, but the display was arranged in such a way as to give credit to Edison. Needless to say, I was astounded.

Further investigation revealed we were not the first group pressuring Dr. Finn to recognize Tesla. A congressman several years earlier had chided Finn to create a Tesla display. The display prepared by Dr. Finn consisted of a small glass showcase housing a few insignificant personal artifacts. The showcase was placed in a darkened hallway next to the Men's Room; their main gallery was devoted to an elaborate Edison display.

By this time I had thoroughly read Dr. Finn's book, LIGHTING A REVOLUTION, given me by Finn's assistant when I visited the museum. The section entitled, "The Beginning of the Electrical Age," named 43 contributors to the science of electricity, but Tesla's name was omitted! This was particularly disturbing because, in describing the Niagara Falls Power Project Dr. Finn alluded to Edison's genius that made the Niagara project possible: "When the Niagara Falls power station began operating in 1895, it signaled the final major act in the revolutionary drama that began in Menlo Park in the fall of 1879." In other words, in 1879 Edison invented DC electricity! Then, in 1895, he invented AC and harnessed Niagara Falls! The historical truth of this episode is that Edison fought the rise of AC and played no role whatsoever in the Niagara Falls power project. An acknowledgement on the inside of the book revealed that its funding was made possible by the Thomas Alva Edison Foundation. Is it reasonable to believe that Dr. Finn's depiction of electrical history was highly biased toward Edison because funding for the book came from the Edison Foundation?

In "Science", Vol.245, July - Sept. 1989, p.768, a book review entitled "The Real Edison, The Making of an Inventor," by Reese V. Jenkins et al., Eds. Johns Hopkins University Press, Baltimore, 1989, carries a most revealing observation: "This volume renders untenable the popular view of Edison as a Merlinesque figure who wrestles with mysterious forces and single-handedly conjures up stunning inventions." It seems I have an ally at Johns Hopkins University.

Before being married, my daughter taught Physics and Chemistry for ten years at Eastern Michigan University. I asked her to involve her physics students in my campaign. They wrote numerous letters to the Secretary of the Smithsonian besieging him to accept our bust. All responses were negative because the Secretary was determined to back Finn's personal bias toward Edison, amounting to the virtual exclusion of Tesla.

One of my students suggested we make promotional T-shirts. We engaged an artist to create a six-color likeness of Tesla including several symbolic figures representing highlights in his life. The artwork and silk-screening was complex, and we could find only one firm in Ann Arbor able to do the work. The end product was magnificent! We sold hundreds of shirts in almost every state. the proceeds went into our school's tax-free account.

My next move was to write U.S. Senator Carl Levin from Michigan; we asked if he could intercede on our behalf to the Smithsonian. Several letters were exchanged, but they had no apparent effect other than to annoy Smithsonian officials. Then in one final effort, Senator Levin, on July 10, 1990, stood on the floor of the U.S. Senate and delivered a scorching account of the Smithsonian's historical debacle regarding Tesla: (Congressional Record, Vol. 136, No. 86): "...Nikola Tesla has not been granted his proper place in history. In the Smithsonian Institution, for example, Mr. Edison's inventions are justifiably well represented. However, although the museum has included Mr. Tesla's alternating current generators in their exhibit, no mention is made of Mr. Tesla. In fact, the generator is included as part of the Edison exhibit..."

At this time several Michigan newspaper articles telling our story appeared in Detroit and locally. Later, an Associated Press writer put the story in newspapers throughout the country. I even received one report that someone saw our article in a Hong Kong newspaper. All this publicity had no effect on Smithsonian policy. Finally, the Secretary promised to have Dr. Finn create a significant Tesla display in 1993. Their plan was to commemorate the Hundredth Anniversary of the Columbian Exposition in Chicago. This was the first World's Fair brilliantly lit by electricity--the same AC that powers the world today made practical by Tesla's genius.

I sensed the Secretary was only buying time and that some excuse would be found not to give Tesla his due. It was not long until my apprehension became reality. Dr. Finn was relying on obtaining a few Tesla artifacts held in a Yugoslav museum. Soon thereafter the war in Yugoslavia erupted and Dr. Finn used the war as an excuse not to create a Tesla display. The Yugoslavian museum items he deemed so important for his proposed display were only insignificant artifacts...a suit of clothing, a cane, a pair of shoes, a suitcase, etc. Obviously, the public would not relate these items to Tesla's greatest contributions--his rotating magnetic field principle and his basic four-tuned circuits forming the essential element of radio. Dr. Finn's plan is to explain Tesla as an eccentric loner who remained outside the arena of mainstream academe and industry. Would not science history be better served if he highlighted and paid tribute to Tesla's greatest contributions? These fundamental discoveries serve as the foundation for our two most important industries--power and communication.

It was Hans Christian Oersted discovering electromagnetism in 1820, followed by Michael Faraday making the first electromagnetic generator in

1831, who really opened the age of electric power. Tesla's rotating magnetic field principle indeed 'signaled the final major act in the revolutionary drama'...but that drama began with Oersted and Faraday, not with Edison at Menlo Park!

Veterans' groups recently severely criticized the Smithsonian, and rightly so, for their wrongful depiction of history regarding the bombing of Hiroshima and Nagasaki. The outcry was so loud that Smithsonian officials had to modify their Enola Gay display. I see little difference in the historical bias the Smithsonian shows against Tesla and the nonsensical bias they preferred regarding the Hiroshima and Nagasaki bombing. Of course the main difference in these disputes is that Tesla does not have large segments of the general public complaining to Smithsonian officials.

Unfortunately the Smithsonian is a political organization, much the same as many other groups in Washington--they blow in the same direction as the wind. I believe the wind has been blowing for too long in Edison's direction, particularly in regard to his entrepreneurial experiment with DC power. I want to see credit given where credit is due...to Tesla for his polyphase AC system...the system that works. We should not be celebrating a businessman's failure in the power industry.

The History of Technology faction that emerged in recent years has subverted our national museum and converted it into an advertising medium promoting specific industries that support them financially. Also, the group that has assumed power at the Smithsonian is methodically dismantling and converting our national museum into an instrument for social change. What was once a tribute to America is gradually being transformed into an assault on American history. I strongly suspect these issues were not even imagined by Mr. James Smithson, a genteel and erudite Englishman, when he bequeathed his estate to the government of the United States to establish the Smithsonian "for the increase and diffusion of knowledge." Smithson was a scientist and man of letters, not a social activist. He undoubtedly envisioned a museum that celebrated creativity (instead of entrepreneurship), and tasteful displays depicting an accurate account of our society's accomplishments (instead of focusing on its problems).

OERSTED, FARADAY, AND TESLA:

Their Discoveries Created our Worldwide System of Electric Power

The discoveries made by these three Giants of Electricity, perhaps more than any other scientific breakthroughs since 1888, are responsible for advancing mankind as no other discovery has since Johann Gutenberg invented the movable type printing press in 1453.

Focusing on three is probably unfair to many others who also made significant contributions in electrical science. Andre-Marie Ampere (1775-1836), a mathematician; and Georg Simon Ohm (1788-1854), a school teacher, made truly significant contributions, establishing the science of electrodynamics and electrical conduction respectively. The properties of electromotive force in Ampere's time were in a constant state of interpretation and revision by many investigators; nevertheless, it was Ampere's brilliant deduction that solved the scientific riddle. Between the years 1820 and 1823 Ampere experimented with current carrying conductors and reasoned that electromotive force is manifested by two kinds of effects: electric tension and electric current; thus, he established the concepts of voltage and current. Three years later Ohm defined the exact relationship between these two entities. Today, Ohm's Law remains the most basic and universally used of all laws in electrical science.

There were many others, too numerous to mention in this short account, who also made advancements. The names of Joseph Henry (1797-1878), James Clerk Maxwell (1831-1879), and Heinrich Hertz (1857-1894) stand out prominently. Henry, an American physicist, experimenting with electromagnets in 1831, discovered self-inductance, a separate phenomenon of electromagnetic induction. Maxwell, a Scottish mathematician genius, established the laws of electrodynamics by formulating four foundational equations defining electromagnetic theory. He concluded that energy could be transmitted by electromagnetic waves at the speed of light. Hertz was convinced that Maxwell's mathematical postulates were correct, so he set out to interpret and prove experimentally Maxwell's work. His experiments proved those postulates...that electromagnetic waves were indeed propagated in air at the speed of light.

In most long and arduous journeys into the unknown, interim breakthroughs are reached long before the ultimate goal is achieved. The journeys traveled by Oersted, Faraday, and Tesla clearly illustrate this reality. In short, these men made those pivotal breakthroughs in electrical research that gave us our worldwide system of electric power.

One thing is certain: the single characteristic all great scientists share is, insatiable curiosity. Perhaps more often than not, luck plays an important role, and a great discovery is made accidentally when the curiosity seeker simply stumbles upon an eternal principle, often doing so without realizing the significance of his discovery.

Thales of Miletus was one of the Seven Wise Men of ancient Greece who lived some 2,500 years ago. In his search for the substance from which everything in nature is made, he accidentally discovered static electricity by rubbing a piece of amber with cloth, resulting in the amber attracting bits of feathers and the pith of plants. The discovery of the lodestone's magnetic attraction is another example of early 'curiosities in nature' that ultimately led to present day knowledge...a world these early curiosity seekers could not possibly envision.

It was not until many hundreds of years later that someone made the next truly significant advancement. In 1453 Johann Gutenberg changed civilization forever with his movable type printing press...a significant factor giving birth to the Renaissance period that swept throughout Europe. The Renaissance is characterized as a rebirth in all forms of learning, and revival of interest in the wisdom of the ancient world of the Near East and Greece. Two names from that period stand out as beacons lighting the way for others to follow: Galileo Galilei (1564-1642), an Italian, and William Gilbert (1544-1603), an English physician. They were the first to reject ancient thinking, which for hundreds of years had dominated man's knowledge of his world. Galileo is called the founder of modern experimental science. It was his bold defiance of ancient teachings that established the beginnings of the scientific method. He made his own telescopes--the largest available--and gathered much new information about the moon, stars, and planets. In addition, his experiments with the pendulum and falling bodies changed forever previously held beliefs, adding greatly to the understanding of physics. In England, Gilbert's scientific study of the magnet culminated in a monumental treatise entitled De Magnete. His quest was to improve the accuracy of the magnetic compass for better navigation on the seas and oceans. What he never realized was that he also laid the foundation of magnetic science, a key element that eventually led to the generation of electricity.

Gradually, scientific investigators entered a period dominated by the study of electrostatics...a curious phenomenon of nature, but still having little or no practical value. The major thrust of scientific investigation during this period--as far back as Gilbert--still centered on improving the magnetic compass, although many independent investigators were working on the properties of conductivity. Charles A. Coulomb (1726-1806) was most prominent during this era because he established the fundamental laws of static electricity, and later made significant advancements in the manufacture of compass needles. Investigators became proficient at generating and instantly discharging static electricity, but they had no way of storing it. E. G. von Kleist and Pieter van Musschenbroeck bridged that gap in the early 18th century with their discovery of the Leyden Jar, a capacitive device that could store static electricity for discharge later. It became a novelty item for royalty who took pleasure in shocking unsuspecting victims...but all in good fun. There was still not a hint of where electrical investigation would eventually take mankind.

Then in 1800, Alessandro Volta (1745-1827), made the first electrochemical cell and battery capable of producing continuous electric current. His inspiration came from Luigi Galvani (1737-1798), a physician who had been conducting experiments with frog legs hung on brass hooks. The legs convulsed when he touched a piece of iron to the framework. Galvani proposed a theory of "animal electricity" as the reason the frog legs had muscle spasms. Volta disproved this theory, stating accurately that the frog legs convulsed as a result of their being in contact with two different metals. His metal theory intrigued him, so he conducted numerous experiments.

Eventually, Volta created a chemical cell capable of producing a continuous electric current. He assembled zinc disks alternately with silver disks, separated by pasteboard soaked in brine solution, and called it an "electric pile." No longer was static electricity the only form of electricity known to man. Volta's continuous current cell was indeed a milestone in the annals of discovery. It was Volta's chemical cell that truly put electricity on the move, and today the battery is still an important, although minor, source of electrical energy.

The next exciting event took place in 1820, in the classroom of a Danish professor named Hans Christian Oersted (1777-1851). Oersted was conducting an experiment with one of his students, showing him how a wire could be heated when it is connected to a voltaic pile. Oersted had neglected to clear the table after his previous experiment, and a magnetic compass remained near the wire. When the connection was made to the voltaic pile, the compass needle turned and pointed toward the wire. At first, Oersted could not believe what he had seen, but ultimately he realized he had discovered something new...electricity and magnetism were interrelated. He named this new force in nature electromagnetism. Without realizing it, Oersted had discovered the magic doorway which would ultimately lead to the age of electricity.

Michael Faraday (1791-1867), made the next giant step on the long road leading to modern day electric power. The impetus that set him on his series of epoch-making experiments was news of Oersted's discovery of electromagnetism. Faraday reasoned that if electricity produces magnetism, then why shouldn't magnetism produce electricity? Finally, in 1831, his experiments revealed a great truth...electricity could indeed be produced by magnetism. The critical component of his discovery, however, was that magnetism must be accompanied by motion. Unlike Oersted, who accidentally stumbled upon his discovery of electromagnetism, Michael Faraday worked diligently toward his goal until he achieved it. If Oersted discovered the magic doorway that would lead to the age of electric power, it was Faraday who unlocked the door. His public wondered what use could possibly come from producing a small current by moving a magnet near a length of wire--the Genie still needed to be tamed to become man's tireless servant. Faraday understood the far-reaching possibilities and is said to have replied: "What is the use of a new-born baby?" Further example of Faraday's wit is in English folklore. The Prime Minister is said to have asked him what use could be made of his discoveries. Faraday allegedly responded, "Someday it might be possible to tax them."

For the next 51 years man struggled to increase the electrical output of Faraday's embryonic generator. His genius had shown the way; now it was up to engineers to make progressive refinements of Faraday's discovery. Volta's battery had been used initially for electroplating metals, but they were large and very expensive. Gradually, engineering advancements in magneto-electric generators became available. Besides their use in the electroplating industry, these first generators were also used for powering arc lamps, lighthouses, and naval vessels. Arc lamps required DC, so a commutator was necessary for rectifying naturally occurring AC. By 1872 the DC generator had reached its peak of refinement, and DC motors had also come into limited use. Even so, they proved to be inefficient and troublesome, creating sparks and requiring frequent maintenance. In addition, DC power was inherently inferior because of its I square R voltage loss, and could not be sent a distance greater than 1/2 mile from the generating station. DC also required inordinately large cables to transmit the current, making it very expensive. Nevertheless, this was the only path man knew to follow.

In 1879 the Menlo Park group, headed by Thomas Edison, invented the first practical incandescent lamp...a much needed improvement over the power hungry arc lamp. Edison, whose thirst for entrepreneurial conquests exceeded (or perhaps equaled) his popularity as an inventor, saw the opportunity to capitalize on the new incandescent lamp by using existing DC technology as a power source. He attracted investors to fund the construction of a power station in New York. Soon, more DC power stations were built in America and abroad. Their existence proved to be a short lived anomaly or evolutionary trend, doomed to extinction because of their inherent inefficiency. DC power, even in its refined state, barely cracked the door to the awesome power contained in Faraday's Genie.

Harnessing alternating current was deemed impossible by leading authorities, including the physics professor of a young engineering student in Austria named Nikola Tesla. He had exactly the opposite idea of his professor and set out to prove him wrong--that AC could indeed be harnessed. Professor Poeschl and the classmates laughed at Tesla, but he was undaunted. For two years he went into a state of self-imposed exile, devoting his entire energy to solving the AC conundrum. He sensed that the answer was hidden in his mind and would eventually come forth, but the mental anguish he suffered during his search nearly killed him, so strong was his resolve. Then in 1882 the solution came, suddenly, in a blinding flash, as he recited poetry while walking in a park. In that instant his brilliant mind conceived perhaps the most beautiful and ingenious scientific creation since the invention of the wheel; he called it the rotating magnetic field.

Tesla's understanding of the awesome power contained in Faraday's Genie was clearer than any of his contemporaries could imagine. The problem of how to harness that energy to do the work of giants had dogged him for more than two years, and now he had discovered the final solution to make alternating current man's servant! The scientific community had always regarded AC equal to a perpetual motion concept, utterly ridiculous! During Tesla's sudden burst of brilliance on that day in the park he not only conceived the rotating magnetic field, but an entire system of polyphase AC which has remained unchanged in principle to this day. Energy in the form of electric power could now be used in virtually unlimited amounts anywhere. His rotating magnetic field discovery, equal in importance to Oersted and Faraday's pioneering achievements, is a principle of unfathomable beauty that will live forever. It stands today as the foundation upon which our entire world operates. Indeed, the power of Faraday's Genie, captured by Tesla, was and still is the most important scientific discovery in more than 500 years. Tesla had not only opened Faraday's magic door; his polyphase system literally blew it off its hinges and the entire wall with it...so great was the power of Faraday's Genie!

Today, the world owes Oersted, Faraday, and Tesla a tremendous debt of gratitude, for it was their monumental discoveries in electrical science that gave us the power to run our factories, mills, schools, hospitals, research centers, stores, and homes.

Technological advances are happening at an ever increasing rate, and we seem always to be looking forward, which is good, but would it not also be uplifting to look back and acknowledge, even pay homage to, the pioneering spirit of those whose efforts laid the foundation for all the technological advantages we enjoy today? Our school books, teachers, and professors seldom mention these pioneers, if indeed they know anything of their existence. Further, many historians and book writers have elevated the names of entrepreneurs and technologists for discoveries made by early pioneers, and if we are not more mindful, our historical heritage will be lost forever.

Welcome to the Tesla Memorial Society of New York

John Wagner - Placing Nikola Tesla Busts at various major Universities in the USA.
John Wagner, school teacher from Ann Arbor, Michigan, mobilized his third grade classes to raise thousands of dollars in funds to create commemorative bronze busts of Tesla to be placed at major American Universities. Many busts are already placed at many Universities.

Mr. Wagner initiated the letter writing campaign to restore Nikola Tesla to his rightful place in United States historical records.

John Wagner is selling Tesla T-shirts to product some funds to buy Tesla bronze busts to be placed at major American Universities. The Tesla Memorial Society of New York is grateful to Mr. Wagner for his campaign to keep Tesla's name and memory alive.

Please contact Mr. Wagner if you wish to buy Tesla T-shirts, by doing so you will help in the purchasing of Tesla bronze busts which will be placed at major American Universities.

Please contact:

John W. Wagner

3890 Tubbs Road

Ann Arbor, Michigan 48103-9437

< http://www.concentric.net/~jwwagner >

Reference[xxv] Article - Nikola Tesla - Forgotten American Scientist. Extracted from the Internet http://www.ntesla.org/ntesla/NT-P.13.html January 18, 2019. Posted in near complete entirety in honor of both the publisher John W. Wagner, school teacher and his commemorative efforts of American inventor recognition of Nikola Tesla

Many of Tesla's inventions were classified

When Tesla died in 1943, during World War II, the Office of Alien Property took his belongings, Alcorn said. Most of his things were later released to his family, and many ended up in the Tesla Museum in Belgrade, which opened in the 1950s. But some of Tesla's papers are still classified by the U.S. government.

"I know people have requested things through the Freedom of Information Act, and they are released heavily redacted," said Alcorn.

As a result of the years of secrecy, many people have speculated about what fantastic inventions might have been suppressed, perhaps to keep them out of enemy hands or, more darkly, to perpetuate the status quo. Perhaps supporting the former theory, Tesla had spoken publicly about working on a "death beam." Those who fear the latter theory often point to his work on harvesting the energy in the forces of nature as something that would upset powerful oil companies.

For her part, Alcorn said she is inspired by both the genius and the perseverance of Tesla. "He taught us that when you believe in yourself, work on your goals, and follow through, a lot is possible," she said.

Article titled - 5 Surprising Facts About Nikola Tesla , paragraph 15. Written By Brian Clark Howard whom covers environment, science, tech, and other topics, and is active in social media. He started at National Geographic in 2011. Extracted from the Internet @ https://news.nationalgeographic.com/news/2013/10/131003-nikola-tesla-surprising-facts-statue-museum-science/

Nikola Tesla's 5 Lost Inventions That Threatened The Global Elite

(anonhq.com) Most great inventions fundamentally change the society in which they exist. Since the people at the top of the social structure have more to gain by reinforcing the status quo, they suppress revolutionary technologies favorable to the world but dangerous to their existence.

Engineering genius Nikola Tesla was no exception. Here are some of those technologies, 'they' don't want you to know about Nikola Tesla. (Video below as well!)

Improved Airships

Tesla proposed that electrically-powered airships would transport passengers from New York to London in three hours, traveling eight miles above the ground. He also imagined that airships might draw their power from the very atmosphere, never needing to stop for refueling. Unmanned airships might even be used to transport passengers to a preselected destination or for a remote aerial strike. He was never given credit for his invention. However, today, we have unmanned drones carrying out combat missions, supersonic airplanes that fly at amazing speeds and space shuttle technology that can circle the Earth in the upper atmosphere.

It was long suspected that the FBI literally stole all of his work, research, and inventions that he had in his possession when he died. This rumor has now been confirmed by recent, heavily redacted Freedom of Information Act requests released by the FBI.

Death Ray

Nikola Tesla claimed to have invented a "death beam" which he called Teleforce in the 1930s. The device was capable of generating an intense targeted beam of energy "that could be used to dispose of enemy warplanes, foreign armies, or anything else you'd rather didn't exist".

The so-called "death ray" was never constructed because he believed it would become too easy for counties to destroy each other. Tesla proposed that a nation could "destroy anything approaching within 200 miles... [and] will provide a wall of power" in order to "make any country, large or small, impregnable against armies, airplanes, and other means for attack". He said that efforts had been made to steal the invention. His room had been entered and his papers had been scrutinized, but the thieves, or spies, left empty-handed.

Tesla's Oscillator

In 1898, Tesla claimed he had built and deployed a small oscillating device that, when attached to his office and operating, nearly shook down the building and everything around it. In other words, the device could allegedly simulate earthquakes. Realizing the potential terrors such a device could create, "Tesla said he took a hammer to the oscillator to disable it, instructing his employees to claim ignorance to the cause of the tremors if asked". Some theorists believe the government continues to use Tesla's research in places like the HAARP facility in Alaska.

Free Electricity System

With funding from JP Morgan, Tesla designed and built Wardenclyffe Tower, a gigantic wireless transmission station, in New York in 1901-1902. Morgan thought the Wardenclyffe Tower could provide wireless communication across the world. However, Tesla had other plans.

Tesla intended to transmit messages, telephony and even facsimile images across the Atlantic to England and to ships at sea based on his theories of using the Earth to conduct the signals. If the project worked, anyone could have electricity by simply sticking a rode into the ground. Unfortunately, free electricity is not profitable. And this system could be incredibly dangerous for the global elite because it could profoundly change the energy industry. Imagine how different the world would be if society didn't need oil and coal to function? Could the great world powers maintain control? Morgan refused to fund the changes. The project was abandoned in 1906 and never became operational.

The Flying Saucer

In 1911, Nikola Tesla told The New York Herald that he was working on an anti-gravity "flying machine".

"My flying machine will have neither wings nor propellers. You might see it on the ground and you would never guess that it was a flying machine. Yet it will be able to move at will through the air in any direction with perfect safety, at higher speeds than have yet been reached, regardless of weather and oblivious of "holes in the air" or downward currents. It will ascend in such currents if desired. It can remain absolutely stationary in the air, even in a wind, for great length of time. Its lifting power will not depend upon any such delicate devices as the bird has to employ, but upon positive mechanical action."

Tesla's flying saucer was powered by free energy system at a time when the fledgling aviation and motor car industry depended on oil and petroleum. His invention met the same fate as his free energy system.

Reference[xxvi] Article - Nikola Tesla's 5 Lost Inventions That Threatened The Global Elite. Published January 2, 2017 by Collective Evolution. Extracted from the Internet on January 9, 2019 @ https://www.collective-evolution.com/2017/01/02/nikola-teslas-5-lost-inventions-that-threatened-the-global-elite/

Tesla and EMF

Military interest into the weaponization of the electromagnetic spectrum has a long history, based on the theoretical work of Nikola Tesla. Radar, in its early inception, was seen not only as a means of tracking the position and speed of enemy targets, but as a potential weapon in its own right. There are very real problems however with overcoming the normal decrease in effect of an electromagnetic field over distance. This effect is a natural function of the laws of physics and applies to both electrical and magnetic fields[39].

In short, the strength of a field drops off in inverse proportion to the distance of the target from the source. Without a means of concentrating and directing a beam of energy across long distances, any effect that an EMF weapon may have would be limited to its immediate vicinity. From 1900 until his death in 1943, Nikola Tesla worked to develop just such a weapon.

In a letter to the New York Times editor in 1908 Tesla wrote,

"When I spoke of future warfare I meant that it should be conducted by direct application of electrical waves without the use of aerial engines or other implements of destruction... What I said in regard to the greatest achievement of the man of science whose mind is bent upon the mastery of the physical universe, was nothing more than what I stated in one of my unpublished addresses, from which I quote:

"According to an adopted theory, every ponderable atom is differentiated from a tenuous fluid, filling all space merely by spinning motion, as a whirl of water in a calm lake. By being set in movement this fluid, the ether, becomes gross matter. Its movement arrested, the primary substance reverts to its normal state. It appears, then, possible for man through harnessed energy of the medium and suitable agencies for starting and stopping ether whirls to cause matter to form and disappear. At his command, almost without effort on his part, old worlds would vanish and new ones would spring into being.

"He could alter the size of this planet, control its seasons, adjust its distance from the sun, guide it on its eternal journey along any path he might choose, through the depths of the universe. He could make planets collide and produce his suns and stars, his heat and light; he could originate life in all its infinite forms. To cause at will the birth and death of matter would be man's grandest deed, which would give him the mastery of physical creation, make him fulfill his ultimate destiny."[40]

Tesla made several claims during the latter years of his life, published by the New York Times in what became an annual event. His theory of the hidden nature of our universe supplants those of many of his contemporaries in that he was able to infer a multidimensional model of the universe that is only now being investigated through the theoretical mathematics of our leading physicists.

Tesla also developed means of remotely controlling aircraft as early as 1915, foreshadowing the Unmanned Ariel Vehicles (UAVs) of today's battlefields. In 1934 Tesla offers to build a "Death Ray" that would make the power of an opponent's air force obsolete. This was one of the earliest recorded statements regarding directed energy weapons.[42] Tesla's offer to build this device for the US government for a bargain price, but with many caveats, was refused by officials who, preferred instead to pump money into the new Army Air Corp, which in turn gave rise to the military aviation complex that we have today.

Before the war the airline industry was not a major part of the economic life of the nation.[44] With huge wartime contracts, however, corporations such as Hughes, McDonnell Douglas, Lockheed, and Northrop quickly grew in power commensurate with the financial bonanza that was unearthed in the battlefields of Europe and the Pacific. These companies formed the core of the "military-industrial complex."

Their investors and managers began to consolidate their clout in political circles to keep the nation on a wartime economic footing, a simple and vastly powerful weapon that would make aircraft, bombs, missiles and attendant industries irrelevant would certainly be seen as a direct threat to the growing power of military arsenal.

Instead, a "black budget" program was put into motion, which exploited the work of Robert Oppenheimer, Albert Einstein and others. The Manhattan Project, developed by the DOD in 1942, generated a vastly destructive weapon that required a well-established and unbelievably expensive aerospace industry, along with unprecedented levels of secrecy and autonomy from Congress and the public.

The US government also ignored Tesla's offer to produce a "city killing machine," which was composed of an electromagnetic shield and a wireless torpedo. Tesla made several proposals during the 1930s, none of which received funding. Among Tesla's claims, published annually on his birthday in the New York Times, were methods of harnessing the power of the sun to electrify the earth and provide free electrical power to anybody, anywhere.

Tesla did, however, conceive of at least one device that became a major part of our nation's arsenal – radar. As early as 1917 he published his theory and developed the first prototype in 1934. It is from the basis of this technology that future research into weaponizing the electromagnetic spectrum proceeded. At the same time Tesla was working on methods of transmitting and receiving communication signals through interplanetary space and reading the images on a sleeping person's retina (by extension mind reading). His prediction that future wars would be fought with electromagnetic means foreshadowed the rise of electronic warfare and the non-lethal weapons technology being deployed today.

At first glance, it would seem probable that the military had taken over the management of Tesla's material. In fact, a number of projects related to his life's work were in development. For instance, the building of beam weapons at Wright Patterson Air Force Base under the code name "Project Nick"[48] headed by Brigadier General L.C. Craigie. This project was however, cancelled due to an apparent lack of understanding of Tesla's means of transmitting high-energy waves without a loss of power over great distances.

Defense Advanced Research Projects Agency (DARPA) began another project in 1958 codenamed "Seesaw" at Lawrence Livermore Labs[49] aimed at combating reported Soviet advances in electromagnetic weapons and defenses, advances that many believe came about after 1952 when the bulk of Tesla's research and personal effects were turned over to his nephew, Sava Kosanovic, who promptly whisked them away to Yugoslavia. Eight years later Soviet Premier Nikita Khrushchev would state that, "A new and fantastic weapon is in the hatching stage,"[50] horrifying many and prompting calls for more effective means of using EMF, espionage and counter-espionage.

On February 9, 1981, the office of the Undersecretary of Defense Research and Engineering department sent a letter to the FBI that requested the papers of Tesla, stating, "We believe that certain of Tesla's papers may contain basic principles which would be of considerable value to certain ongoing research within the DOD. It would be very helpful to have access to these papers. The letter was signed by Lt. Col. Allan J. Mclaren, an R.O.T.C. graduate from M.I.T. in 1960, who later went on to become a project director with Lockheed Martin Space Systems from which he retired in 2003.[51]

This section of his memo to the FBI was not declassified until 1993. In response, the FBI issued the same response as to all of the other inquiries with one exception, this time they identified who it was that examined the stored effects; it was the Office of Scientific Research and Development from MIT, a breeding ground of CIA.[52]

What they may have been looking for had likely already been taken, according to a recent PBS special entitled Tesla: Life and Legacy, Tesla's nephew reported that Tesla's most recent journal was missing from the bulk of material stored by the OAP.[53] In recent years high profile projects such as the High Altitude Auroral Project ("HAARP"), the Strategic Defense Initiative ("Star Wars"), and many of the devices promoted by proponents of "Non-Lethal Weapons" have Tesla's intellectual fingerprints all over them.[54]

Reference[xxvii] Article – Tesla and EMF Has multiple references [51] See Tesla's FBI files at the FBI FOIA site located at https://vault.fbi.gov/nikola-tesla.

[52] "Tesla: Life and Legacy, Missing Papers," archived at PBS: http://www.pbs.org/tesla/ll/ll_mispapers.html.

[53] Tesla: Master of Lightning."" PBS documentary Dec. 12th, 2000.

[54] Box #8 of Declassified CIA documents pertaining to MKULTRA contains the following fragment: The Application of Tesla's Technology in Today's World. Obtain, online, through the National Security Archives at http://www.gwu.edu/~nsarchiv/

Nikola Tesla - Magnifying Transmitter

Cloud born Electric Wavelets To Encircle the Globe. This is Nicola Tesla's Latest Dream, and the Long Island Hamlet of Wardenclyffe Marvels Thereat

To gather in the latent electricity in the clouds and with the globe itself as a medium of transmission to convey telegraphic messages, power for commercial purposes, or even the sound of the human voice to the utmost confines of the earth is the latest dream of Nikola Tesla. In an article which appeared recently in "The Electrical World," Mr. Tesla explains the theories on which the world telegraphy system is founded and what he expects to accomplish by it.

His plans involve the establishment of stations for the transmission of messages and power, "preferably near important centers of civilization." Oddly enough, what Mr. Tesla proudly designates as the first of his commercial "world telegraphy" stations has been established at Wardenclyffe, L.I., which is not in any sense an important "center of civilization," but a place described by train hands of the Long Island Railroad as a way station where "a passenger alights occasionally."

Nikola Tesla Magnifying Transmitter.

The transmitting station is an octagonal tower, pyramidal in shape, and some 187 feet in height. It consists of huge wooden stilts, heavily braced, and reinforced, and surmounted by a cupola of interlaced steel wires, bent so as to form an arc. In the cupola there is a wooden platform occupying its entire width. Mr. Tesla began work on his transmitting station about eighteen months ago.

When he first came there, and it was understood that J. Pierpont Morgan had become interested in his odd enterprise and furnished him with financial assistance, a thrill of vague expectancy ran through the little settlement, The Wardenclyffe Land Company, which owns practically all the available ground in the vicinity, gave the inventor a free grant of some 175 acres of fine land, and then settled down to wait for the day when Wardenclyffe would become the center of the universe.

Some of the farmers who come to Wardenclyffe to send their products to this city look at Mr. Tesla's tower, which is situated directly opposite the railroad station, and shake their heads sadly. They are inclined to take a skeptical view regarding the feasibility of the wireless "world telegraphy" idea, but yet Tesla's transmitting tower as it stands in lonely grandeur and boldly silhouetted against the sky on a wide clearing on the concession is a source or great satisfaction and of some mystification to them all.

"It is a mighty fine tower," said one food farmer to a visitor last week. "The breeze up there is something grand on a Summer evening, and you can see the Sound and all the steamers that go by. We are tired, though, trying to figure out why he put it here instead of at Coney Island. " While the tower itself is very "stagy" and picturesque, it is the wonders that are supposed to be hidden in the earth underneath it that excite the curiosity of the population in the little settlement.

In the center of the wide concrete platform which serves as a base for the structure there is a wooden affair very much like the companionway on an ocean steamer. The tower and the enclosure in which it has been built are being carefully guarded these days, and no one except Mr. Tesla's own men are allowed to approach it. Only they have been allowed as much as the briefest peep down the companionway. Mr. Scherff, the private secretary of the inventor, told an inquirer that the companionway led to a small drainage passage built for the purpose of keeping the ground about the tower dry.

But such of the villagers as saw the tower constructed tell a different story. They declare that it leads to a well-like excavation as deep as the tower is high with walls of mason work and a circular stairway leading to the bottom.

Nikola Tesla Magnifying Transmitter 2.

From there, they say, tunnels have been built in all directions, until the entire ground below the little plain on which the tower is raised has been honeycombed with subterranean passages.

They tell with awe how Mr. Tesla, on his weekly visits to Wardenclyffe, spends as much time in the underground passages as he does on the tower or in the handsome laboratory and workshop erected beside it, and where the power plant for the world telegraph has been installed.

No instruments have been installed as yet in the transmitter, nor has Mr. Tesla given any description of what they will be like. But in his article he announces that he will transmit from the tower an electric wave of a total maximum activity of ten million horse power. This, he says, will be possible with a plant of but 100 horse power, by the use of a magnifying transmitter of his own invention and certain artifices which he promises to make known in due course. What he expects to accomplish is summed up in the closing paragraph as follows:

"When the great truth, accidentally revealed and experimentally confirmed, is fully recognized, that this planet, with all its appalling immensity, is to electric currents virtually no more than a small metal ball and that by virtue of this fact many possibilities, each baffling imagination and of incalculable consequence, are rendered absolutely sure of accomplishment; when the first plant is inaugurated and it is shown that a telegraphic message, almost as secret and non-interferable as a thought, can be transmitted to any terrestrial distance, the sound of the human voice, with all its intonations and inflections faithfully and instantly reproduced at any other point of the globe, the energy of a waterfall made available for supplying light, heat or motive power, anywhere...on sea, or land, or high in the air...humanity will be like an ant heap stirred up with a stick. See the excitement coming!"

"Cloud born Electric Wavelets To Encircle the Globe: This Is Nicola Tesla's Latest Dream, and the Long Island Hamlet of Wardenclyffe Marvels Thereat," New York Times, 27 March 1904

Reference[xxviii] Article - Nikola Tesla Magnifying Transmitter, New York Times - March 27, 1904 Extracted from the Internet January 16, 2019 @ https://www.wanttoknow.info/nikolateslamagnifyingtransmitter

Section Six

Energy News

Excerpts of Key Articles in Media

BlackLight's physics-defying promise: Cheap power from water
2008-07-02, CNN Money

http://money.cnn.com/2008/07/01/smallbusiness/blacklight.fsb/index.htm

Imagine being able to convert water into a boundless source of cheap energy. That's what BlackLight Power, a 25-employee firm in Cranbury, N.J., says it can do. The only problem: Most scientists say that company's technology violates the basic laws of physics. Such skepticism doesn't daunt Dr. Randell Mills, a Harvard-trained physician and founder of BlackLight, who recently claimed that he has created a working fuel cell using the world's most pervasive element: the hydrogen found in water. Mills says he has a market-ready product: a fuel cell that produces a chemical reaction to alter hydrogen atoms. The fuel cell releases heat that turns water into steam, which drives electric turbines. The working models in his lab generate 50 kilowatts of electricity - enough to power six or seven houses. But these, Mills says, can be scaled [up] to drive a large, electric power plant. **The inventor claims this electricity will cost less than 2 cents per kilowatt-hour, which compares to a national average of 8.9 cents.** Mills developed the patented cocktail that enables the reaction - a solid fuel made of hydrogen and a sodium hydride catalyst - only a year ago. (He recently posted instructions on the company's Web site, blacklightpower.com). Now that the device is ready for commercialization, he says, BlackLight is negotiating with several utilities and architecture and engineering firms. **The business, Mills says, has attracted $60 million in funding from wealthy individuals, investment firms ... and it is no longer seeking money.** BlackLight's board of directors reads like a Who's Who of finance and energy leaders.

The coming era of unlimited — and free — clean energy

2014-09-19, *Washington Post* blog

http://www.washingtonpost.com/blogs/innovations/wp/2014/09/19/the-coming-era-...

In the 1980s, leading consultants were skeptical about cellular phones. The handsets were heavy, batteries didn't last long, coverage was patchy, and the cost per minute was exorbitant. The experts are saying the same about solar energy now. They say that solar is inefficient, too expensive to install, and unreliable, and will fail without government subsidies. They too are wrong. Solar will be as ubiquitous as cellular phones are. **Futurist Ray Kurzwell**

notes that solar power has been doubling every two years for the past 30 years — as costs have been dropping. He says solar energy is only six doublings — or less than 14 years — away from meeting 100 percent of today's energy needs. By Kurzweil's estimates, inexpensive renewable sources will provide more energy than the world needs in less than 20 years. In places such as Germany, Spain, Portugal, Australia, and the Southwest United States, residential-scale solar production has already reached "grid parity" with average residential electricity prices. In other words, it costs no more in the long term to install solar panels than to buy electricity from utility companies. The prices of solar panels have fallen 75 percent in the past five years alone and will fall much further as the technologies to create them improve and scale of production increases. By 2020, solar energy will be price-competitive with energy generated from fossil fuels on an unsubsidized basis in most parts of the world. Within the next decade, it will cost a fraction of what fossil fuel-based alternatives do. Despite the skepticism of experts and criticism by naysayers, there is little doubt that we are heading into an era of unlimited and almost free clean energy.

Note: This article also points out how some big energy companies and the Koch brothers are lobbying to stop alternative technologies from flowering.

Cold Fusion Is Hot Again

2009-04-19, CBS News

http://www.cbsnews.com/stories/2009/04/17/60minutes/main4952167.shtml

Twenty years ago it appeared, for a moment, that all our energy problems could be solved. It was the announcement of cold fusion - nuclear energy like that which powers the sun - but at room temperature on a table top. It promised to be cheap, limitless and clean. Cold fusion would end our dependence on the Middle East and stop those greenhouse gases blamed for global warming. It would change everything. But then, just as quickly as it was announced, it was discredited. So thoroughly, that cold fusion became a catch phrase for junk science. Well, a funny thing happened on the way to oblivion - for many scientists today, cold fusion is hot again. "We can yield the power of nuclear physics on a tabletop. The potential is unlimited. That is the most powerful energy source known to man," researcher Michael McKubre told 60 Minutes correspondent Scott Pelley. McKubre says he has seen that energy more than 50 times in cold fusion experiments he's doing at SRI International, a respected California lab that does extensive work for the government.

McKubre is an electro-chemist who imagines, in 20 years, the creation of a clean nuclear battery. **"For example, a laptop would come pre-charged with all of the energy that you would ever intend to use. You're now decoupled from your charger and the wall socket,"** he explained. The same would go for cars. **"The potential is for an energy source that would run your car for three, four years, for example. And you'd take it in for service every four years and they'd give you a new power supply,"** McKubre told Pelley.

2,757.1 MPG Achieved at 2009 Shell Eco-marathon Americas

2009-04-19, CNBC News

https://web.archive.org/web/20120513163728/http:/www.shell.us/home/content/usa/aboutshell/media_center/news_and_press_releases/2009/2009shellecomarathonamericas_finalresults.html

Distance, not speed, was the goal this weekend on the track at the 2009 Shell Eco-marathon Americas(R), a challenge for students to design, build and test fuel-efficient vehicles that travel the farthest distance using the least amount of fuel. This year, **more than 500 students from North and South America were on hand to stretch the boundaries of fuel efficiency. So who came out on top? The student team from Laval University, with an astonishing 2,757.1 miles per gallon, equivalent to 1,172.2 kilometers per liter, won the grand prize in the "Prototype" category.** And in the "Urban Concept" category - new to the Americas event this year - the team from Mater Dei High School took the grand prize by achieving 433.3 mpg, equivalent to 184.2 km/l. With 44 participating teams at track competition was steep. This year's challenge brought together a number of returning teams determined to beat the 2,843 mpg (1,208 km/l) record set by Mater Dei High School (Evansville, Ind.) in 2008, combined with a number of new teams adding fresh innovation and vehicle designs to the competition. "The Shell Eco-marathon is a platform for students to let their imaginations run wild," said Mark Singer, global project manager for the Shell Eco-marathon. "By encouraging these students to build vehicles with greater energy efficiency, we hope this will help inspire others; and together we can find solutions that will help meet the global energy challenge."

Note: Why so little media attention to this most exciting race for top gas mileage? And if high school students can build a car that gets over 2,500 mpg, what's up with Detroit? Could big business be suppressing, or at the very least ignoring these inspiring inventions?

Taming the Gas-Hogging SUV

2008-06-09, ABC News

http://abcnews.go.com/Technology/GlobalWarming/story?id=5028894

Johnathan Goodwin walks to the back of his auto conversion shop in Wichita, Kan., and lifts up a gas nozzle connected to a huge cube-shaped container. The orange stuff he's pumping is the key to his company's mission: converting the worst gas-gulping SUVs into cleaner, meaner machines. "This is 100 percent canola oil, refined to biodiesel," Goodwin said. His well-maintained shop is a bit like a showroom for that much-maligned symbol of environmental ruin: the Hummer. The silver H-1 – which Goodwin says gets 60 miles per gallon – has already been modified to run on biodiesel, diesel, vegetable oil, gasoline, ethanol, hydrogen, natural gas and propane. On a standard gasoline-to-biodiesel conversion, Goodwin starts by taking a new nine-mile-per-gallon Hummer and removing the original gas engine. In goes an off-the-shelf GM Duramax engine that runs on diesel fuel. A few extra modifications and a tank full of biodiesel later, the Hummer – now boasting 500 horsepower and getting about 20 miles per gallon – is ready for the road. He offers a couple of lower-cost options, including a fuel vaporizer for $1,000 that he says boosts fuel economy by 30 percent, and a $500 software download that reprograms diesel engines to get up to an additional seven miles per gallon. **His work has many wondering why the big automakers can't simply reconfigure their assembly lines to make their own cars run as efficiently as Goodwin does. "I don't know why GM hasn't done it," says Goodwin. "But I can tell you that all the parts that I use for the conversion – 95 percent – are all GM parts. I'm not reinventing anything."**

Researcher sets saltwater on fire

2007-11-14, CNN

http://edition.cnn.com/2007/TECH/science/11/14/saltwater.fire/

Last winter, inventor John Kanzius was already attempting one seemingly impossible feat -- building a machine to cure cancer with radio waves -- when his device inadvertently succeeded in another: He made saltwater catch fire. TV footage of his bizarre discovery has been burning up the blogosphere ever since, drawing crackpots and Ph.D.'s alike into a raging debate. Can water burn? And if so, what good can come of it? Some people gush over the invention's potential for desalinization or cheap energy. Briny seawater, after all, sloshes over most of the planet's surface, and harnessing its heat energy could power all sorts of things. Skeptics say Kanzius's radio generator is sucking up far more energy than it's creating, making it a carnival trick at best. For now, Kanzius is tuning out the hubbub. Diagnosed with leukemia in 2002, he began building his radio-wave blaster the next year, soon after a relapse. If he could seed a person's cancerous cells with nanoscopic metal particles and blast them with radio waves, perhaps he could kill off the cancer while sparing healthy tissue. **The saltwater phenomenon happened by accident when an assistant was bombarding a saline-filled test tube with radio waves and bumped the tube, causing a small flash. Curious, Kanzius struck a match. "The water lit like a propane flame,** "he recalls. "People said, 'It's a crock. Look for hidden electrodes in the water,' " says Penn State University materials scientist Rustum Roy, who visited [Kanzius] in his lab in August after seeing the feat on Google Video. A demo made Roy a believer. "This is discovery science in the best tradition," he says. Meanwhile, researchers at MD Anderson Cancer Center in Houston and the University of Pittsburgh Medical Center have made progress using Kanzius's technology to fight cancer in animals. They published their findings last month in the journal Cancer.

The Prophet of Garbage

2007-03-01, *Popular Science* - March 2007 Issue

http://www.popsci.com/scitech/article/2007-03/prophet-garbage

The Plasma Converter ... can consume nearly any type of waste—from dirty diapers to chemical weapons—by annihilating toxic materials in a process ... called plasma gasification. A 650-volt current passing between two electrodes rips electrons from the air, converting the gas into plasma. The plasma arc is so powerful, it disintegrates trash into its constituent elements by tearing apart molecular bonds. The system is capable of breaking down pretty much anything except nuclear waste. The only by-products are an obsidian-like glass [and] a mixture of primarily hydrogen and carbon monoxide that can be converted into a variety of marketable fuels, including ethanol, natural gas and hydrogen. **Perhaps the most amazing part of the process is that it's self-sustaining. Once the cycle is under way, the 2,200°F syngas is fed**

into a cooling system, generating steam that drives turbines to produce electricity. About two thirds of the power is siphoned off to run the converter; the rest can be used on-site for heating or electricity, or sold back to the utility grid. Even a blackout would not stop the operation of the facility. New York City is already paying an astronomical $90 a ton to get rid of its trash. According to Startech, a few 2,000-ton-per-day plasma-gasification plants could do it for $36. Sell the syngas and surplus electricity, and you'd actually net $15 a ton. But the decision-making bureaucracy can be slow, and it is hamstrung by the politically well-connected waste-disposal industry. Startech isn't the only company using plasma to turn waste into a source of clean energy. A handful of start-ups—Geoplasma, Recovered Energy, PyroGenesis, EnviroArc and Plasco Energy, among others—have entered the market in the past decade.

Note: Why isn't this amazing, proven machine and technology making front page headlines?

A Faith-Based Fuel Initiative

2007-01-30, *New York Times*

http://www.nytimes.com/2007/01/30/opinion/30tue1.html?ex=1327813200&en=8ce942...

In 1975, after the oil embargo, Congress approved the most successful energy-saving measure this country has ever seen: the Corporate Average Fuel Economy system, known as CAFE, which set minimum mileage standards for cars. **Within 10 years, automobile efficiency had virtually doubled, to 27.5 miles per gallon in 1985 from just over 14 miles per gallon in 1976. The mileage standards are still 27.5 m.p.g.** Except for minor tweaks, Congress has refused to raise fuel efficiency requirements or close a gaping loophole that lets S.U.V.'s and pickups be measured by a more lenient standard.

Note: Thank you *New York Times* for pointing out what so few have bothered to mention. In the U.S., it is not the automotive industry that determines fuel mileage standards, but rather Congress. Whenever Congress has raised the mileage standard, industry complies and average mileage increases. When the standards are not raised, average car mileage for new cars stays the same. Yet Congress refused to significantly raise the standards from 1985 to 2012, despite the increasing talk of an energy crisis. Why?

Brazil's alcohol cars hit 2m mark

2006-08-18, BBC News

http://news.bbc.co.uk/2/hi/business/5263384.stm

Brazil's new generation of cars and trucks adapted to run on alcohol has just hit the two-million mark. "Flex-fuel" vehicles, which run on any combination of ethanol and petrol, now make up 77% of the Brazilian market. Brazil has pioneered the use of ethanol derived from sugar-cane as motor fuel. Ethanol-driven cars have been on sale there for 25 years, but they have been enjoying a revival since flex-fuel models first appeared in March 2003. Just 48,200 flex-fuel cars were sold in Brazil in 2003, but the total had reached 1.2 million by the end of last year and had since topped two million, the Brazilian motor manufacturers' association Anfavea said.

Note: With sky-high gasoline prices and the fear of depletion of global oil supplies, why don't such cars exist in the U.S.? Why are the trains of almost every other developed nation far advanced from trains in the U.S.? And why isn't the U.S. media reporting on this important development?

Enron Schemes Caught On Tape

2005-02-03, CBS News

http://www.cbsnews.com/stories/2005/02/03/eveningnews/main671618.shtml

During the West Coast Power crisis homes went dark and streetlights were out ... causing injuries and accidents. But the danger didn't stop Enron's energy traders from having a good laugh. CBS ... reports on the Enron scheme, as caught on new audio tape. The traders and plant operator laugh and plot in a display that seems to prove the theory that years before the energy crisis, Enron manipulated markets. "They had to do a rolling blackout through the town and there was a red light there he didn't see," one Enron trader says on tape. "That's beautiful," a second voice responds. **Enron secretly shut power plants down so they could cause, and then cash in on, the crisis.**

Enron also pulled power out of states like California, causing emergency conditions to worsen. "Sorry California," an Enron trader says. "I'm bringing all our power out of state today." Plant operators were coached on how to lie to officials. "We want you guys to get a little creative..." one voice says on the tape, "and come up with a reason to go down. Just call 'em, Hey guys...we're coming down." The plant operator replies, "OK, so we're just coming' down for some maintenance?" "Right," the trader says. "And that's cool?" the plant operator asks. "Hopefully," the trader responds, to which the men are heard laughing. Enron also pulled power out of states like California, causing emergency conditions to worsen. The "shut downs" and "pull outs" triggered sky high power prices. "We're just making money hand over fist!" one voice is heard saying on the tape. And when states complained, the guys at Enron seemed to have a response. "Get a f****** clue," one says. "Yeah," another chimes in. "Leave us alone. Let us make a little bit of money."

Tsunami bomb - NZ's devastating war secret

1999-09-25, *New Zealand Herald* (New Zealand's leading newspaper)

http://www.nzherald.co.nz/nz/news/article.cfm?c_id=1&objectid=14727

Top-secret wartime experiments were conducted off the coast of Auckland to perfect a tidal wave bomb, declassified files reveal. United States defense chiefs said that if the project had been completed before the end of the war it could have played a role as effective as that of the atom bomb. Details of the tsunami bomb, known as Project Seal, are contained in 53-year-old documents released by the Ministry of Foreign Affairs and Trade. Papers stamped "top secret" show the US and British military were eager for Seal to be developed in the post-war years too. The experiments involved laying a pattern of explosives underwater to create a tsunami. It is unclear what happened to Project Seal once the final report was forwarded to Wellington Defense Headquarters late in the 1940s. The bomb was never tested on a full scale. "Whether it could ever be resurrected ... Under some circumstances I think it could be devastating."

Car achieves almost 10,000 miles per gallon

1999-07-16, BBC News

http://news.bbc.co.uk/1/hi/sci/tech/395366.stm

A car driven by a 10-year-old and built at a French school has set a new world record for fuel efficiency. **The Microjoule team managed the equivalent of 9,845 miles per gallon while driving for 10 miles around Silverstone race track in the UK.** More than 100 teams competed in the Shell Eco-Marathon. Their one goal was to see how far they can get these amazing machines to travel on a minuscule amount of fuel. While we might be delirious if we managed 40 miles (64 kilometers) to the gallon (4.5 litres) pottering about town in our super minis, these people are not happy until they have seen the milometer click through the thousands. The teams have a choice of petrol or diesel, with solar assistance permitted for the first time this year. A car is allowed three 40-minute runs. It must average at least 15 mph (24 kph) after which the stewards at the meeting calculate the machine's fuel efficiency. "The top fuel teams do about 10 miles, which is six laps on the club circuit at Silverstone," says the event's fuel manager Geoff Houlbrook. "They do that on less than 10 milliliters which is just two teaspoons of fuel." The entries come from all over Europe. Some teams use advanced materials like titanium and carbon fiber. Some of the machines built by schoolchildren are made from parts of old sewing and washing machines. "It's fun but it's also science," says BBC Top Gear presenter and racing driver Tiff Needell. "It's like an experiment with people learning how to save energy."

Note: Some of these amazing vehicles built in 1999 were "built by schoolchildren," yet the auto industry still can't come up with a car that gets 100 mpg? Granted these cars are slow and small, but if they can get almost 10,000 mpg, don't you think similar technology could be used to get at least several hundred mpg in regular cars?

Was Edison Adversary Father Of 'Star Wars'?

1986-08-10, *Chicago Tribune*

http://articles.chicagotribune.com/1986-08-10/news/8602270598_1_nikola-tesla-...

Weren't we taught that radio was invented by an Italian named Guglielmo Marconi? And that the legendary Thomas Alva Edison devised today's electrical power system? "We were taught wrong," said Toby Grotz, president of the International Tesla Society. Two years before Marconi demonstrated his wireless radio transmission, [Nikola Tesla] performed an identical feat at the 1893 World's Fair in Chicago. **On June 21, 1943, in the case of Marconi Wireless Telegraph Co. vs. the United States, the Supreme Court ruled that that Tesla's radio patents had predated those of the Italian genius.** To be sure, Edison invented the incandescent light bulb. But he powered it and all of his other projects with inefficient direct current (DC) electricity. It was Tesla who discovered how to use the far more powerful phased form of alternating current (AC) electricity that is virtually the universal type of electricity employed by modern civilization. There are indications that Tesla also discovered many of the devices ... for the Pentagon's controversial Star Wars antimissile defense system. "Tesla dreamed of supplying limitless amounts of power freely and equally available to all persons on Earth," said Grotz. And he was convinced he could do so by broadcasting electrical power across large distances just as radio transmits far smaller amounts of energy. **[Tesla's] tests ... caused lights to burn as much as 26 miles away, according to news reports of the time.**

Note: Tesla was written out of history texts likely because he advocated providing methods for extremely cheap electricity available to everyone. He successfully transmitted electricity through the air to lights 26 miles away. Yet the rich energy power brokers of his time could not stand for this. Only the little known Supreme Court ruling mentioned above restored his claim as original inventor of the radio.

Grid parity: Why electric utilities should struggle to sleep at night

2014-03-25, *Washington Post* blog

http://www.washingtonpost.com/blogs/innovations/wp/2014/03/25/grid-parity-why...

What's good news for those concerned with climate change, and bad news for electric utilities? That's grid parity. It exists when an alternative energy source generates electricity at a cost matching the price of power from the electric grid. As grid parity becomes increasingly common, renewable energy could transform our world and slow the effects of climate change. Advances in solar panels and battery storage will make it more realistic for consumers to dump their electric utility, and power their homes through solar energy. **A 2013 Deutsche Bank report said that 10 states are currently at grid parity: Arizona, California, Connecticut, Hawaii, Nevada, New Hampshire, New Jersey, New Mexico, New York and Vermont. Germany, Spain, Portugal and Australia have reached grid parity. This shift has benefited from a dramatic drop in the price of solar panels, which dropped 97.2 percent from 1975 to 2012.** As solar energy gets cheaper, traditional electric utilities are doing the opposite. The cost of maintaining the electric grid has gotten more expensive, but reliability hasn't improved. If customers leave electric utilities, it starts a downward spiral. Fewer customers will mean higher rates, which encourages remaining customers to jump ship for a solar-battery system. Energy upstarts are led by forward thinkers with disruptive track records and eyes on society's big problems. NRG Energy chief executive David Crane ... highlighted the climate change concerns in a recent letter to shareholders: "The day is coming when our children sit us down ... look us straight in the eye, with an acute sense of betrayal and disappointment in theirs, and whisper to us, 'You knew... and you didn't do anything about it. Why?'"

Miracle material graphene can distil booze, says study

2012-01-27, BBC News

http://www.bbc.co.uk/news/science-environment-16747208

Membranes based on the "miracle material" graphene can be used to distil alcohol, according to a new study in *Science* [magazine]. An international team created the membrane from graphene oxide - a chemical derivative of graphene. They have shown that the membrane blocks the passage of several gases and liquids, but let's water through. This joins a long list of fascinating and unusual properties associated with graphene and its derivatives. Graphene ... is a flat layer of carbon atoms tightly packed into a two-dimensional honeycomb arrangement. Because it is so thin, it is also practically transparent. **As a conductor of electricity, it performs as well as copper; and as a conductor of heat, it outperforms all other known materials.** The unusual electronic, mechanical and chemical properties of graphene at the molecular scale promise numerous applications. **Andrei Geim and Konstantin Novoselov from the University of Manchester were awarded 2010's Nobel Prize in physics for their discovery [of graphene]**. Geim and others have now developed a laminate made from thin sheets of graphene oxide. These films were hundreds of times thinner than a human hair but remained strong, flexible and easy to handle. In another study in Science journal, a different team reports the development of a membrane based on diamond-like carbon. This membrane has unique pore sizes that allow for the ultra-fast passage of oil through it. One expert said it could potentially be used for filtering toxic contaminants out of water or for purifying industrial chemicals.

Abiotic Oil: A Theory Worth Exploring

2011-09-14, *US News & World Report* magazine

http://www.usnews.com/opinion/blogs/energy-intelligence/2011/09/14/abiotic-oi...

[Abiotic oil theorists] hold that oil can be derived from hydrocarbons that existed eons ago in massive pools deep within the earth's core. That source of hydrocarbons seeps up through the earth's layers and slowly replenishes oil sources. In other words, it turns the fossil-fuel paradigm upside down. Thomas Gold, a respected astronomer and professor emeritus at Cornell University in Ithaca, NY, has held for years that **oil is actually a renewable, primordial syrup continually manufactured by the Earth under ultra-hot conditions and tremendous pressures. As this substance migrates toward the surface, it is attacked by bacteria, making it appear to have an organic origin dating back to the dinosaurs**, he says. That ... raises the tantalizing possibility that oil may not be the limited resource it is assumed to be. In 2008 ... a group of Russian and Ukrainian scientists [said] that oil and gas don't come from fossils; they're synthesized deep within the earth's mantle by heat, pressure, and other purely chemical means, before gradually rising to the surface. The idea that oil comes from fossils "is a myth" that needs changing according to petroleum engineer Vladimir Kutcherov, speaking at the Royal Institute of Technology in Sweden. "All kinds of rocks could have oil and gas deposits." Alexander Kitchka of the Ukrainian National Academy of Sciences estimates that 60 percent of the content of all oil is abiotic in origin and not from fossil fuels.

Blacklight Power bolsters its impossible claims of a new renewable energy source

2008-10-21, *New York Times*

http://www.nytimes.com/external/venturebeat/2008/10/21/21venturebeat-blacklig...

Ask nearly any physicist if it's possible for a hydrogen atom to enter a lower energy state than the ground, or resting, state they hold in nature, and you're likely to get an unequivocal "no". But a tiny company in New Jersey called Blacklight Power has been disputing that assumption for over a decade, and of late, making gad-fly claims that its founder says will overturn the accepted scientific order. Blacklight's claims have a special significance: If they're true, there's a source of cheap, clean energy that can be easily tapped anywhere in the world. Blacklight is now saying that it has physical proof of its energy generator, verified by an independent university lab. Its "hydrino" theory isn't put forth by a single crackpot; instead, the company employs a good handful of high-level scientists who would presumably rebel if the idea was totally false. It has also taken over $60 million in venture funding. Despite a hearty rejection by the scientific mainstream, and being ignored for years on end, its founder, Randell Mills, has plugged on. Now an engineering team at Rowan University ...has come forward with results from its own tests of the Blacklight process. **Tests conducted in sealed chambers, and measured with a device called a calorimeter, show a heat reaction from a substance provided by Blacklight far beyond anything anticipated. "We've been able to regularly reproduce these results and we believe any research lab could do the same,"** Peter Jansson, the faculty member heading the experiments, [said].

Puzzled Researchers Vet BlackLight's Physics-Defying Hydrogen Power

2008-10-21, *New York Times*

http://www.nytimes.com/external/gigaom/2008/10/21/21gigaom-blacklight-validat...

BlackLight Power, the company that has pulled in $60 million for its seemingly physics-defying fuel cell, is back with an announcement about an independent validation of its technology. A team of engineers, headed by Dr. Peter Jansson at Rowan University, have tested BlackLight's prototypes and found that the devices perform as BlackLight claims, ambiguously concluding that "there is a novel reaction of some type causing the large exotherm which is consistently produced." To translate: There's definitely lots of energy being produced. They're just not sure why. **BlackLight says its technology can push an electron closer to the nucleus by way of a catalytic reaction, resulting**

in a huge amount of clean energy. The company describes the reaction as "somewhere between a nuclear and a chemical reaction," but without any of the messy fallout. The team at Rowan tested BlackLight's 1,000- and 50,000-watt reactors over three months and were able to replicate BlackLight's energy claims, saying that the energy produced "cannot be explained by other known sources like combustion or nuclear energy." The company says a complete verification of the whole process will likely happen within a year. BlackLight tells us it is now in the process of licensing its technology to power producers. The company says it has enough capital to get through commercialization and plans to have its reactors in a power plant in the next two years.

Loremo: The 'Low Resistance Mobile'

2008-02-20, MSN

https://web.archive.org/web/20080308132816/http://editorial.autos.msn.com/art...

The idea is deceptively simple. Forget about fancy batteries, regenerative braking, and alternative fuels. Instead, make a car that's elegant in its minimalism and efficiency. **The Loremo's German designers revisited the basics — engine efficiency, low weight, and minimal drag — to create a car that offers fuel-efficiency in the neighborhood of 130 to 150 miles per gallon. The Loremo is likely to dazzle drivers not with its acceleration, but with its ability to drive from New York to L.A. with only three stops at the pump.** Loremo stands for low resistance mobile, and its engineers have stuck obsessively to this idea. By building the car around a 2-cylinder turbodiesel engine, and cutting back on weight, drag, and other excess fat such as side-opening doors, the Loremo puffs out a mere 50 grams of carbon dioxide per kilometer. This is about 40 grams less per kilometer than the tiny diesel smart. According to its creators, this will make the Loremo the most efficient production car ever sold. If the Loremo showed up as a concept on an auto show pedestal, it would certainly garner some attention. But the Loremo is not a car for dreamers; not only will it enter mass production next year, it will sport a base price attainable by mortal motorists: 15,000 euros (about U.S. $22,000). After its 2009 release in Europe, the Loremo will be redesigned to reach the North American market the following year. A $30,000, 3-cylinder GT model will also become available, offering better acceleration (0-60 in roughly 10 seconds, vs. 16 for the base model). Both hybrid and fully electric versions are also in the works.

The New Dawn of Solar

2007-12-01, *Popular Science* magazine

http://www.popsci.com/popsci/flat/bown/2007/green/item_59.html

Imagine a solar panel without the panel. Just a coating, thin as a layer of paint, that takes light and converts it to electricity. From there, you can picture roof shingles with solar cells built inside and window coatings that seem to suck power from the air. Consider solar-powered buildings stretching not just across sunny Southern California, but through China and India and Kenya as well, because even in those countries, going solar will be cheaper than burning coal. **That's the promise of thin-film solar cells: solar power that's ubiquitous because it's cheap. The basic technology has been around for decades, but this year, Silicon Valley–based Nanosolar created the manufacturing technology that could make that promise a reality.** The company produces its PowerSheet solar cells with printing-press-style machines that set down a layer of solar-absorbing nano-ink onto metal sheets as thin as aluminum foil, so the panels can be made for about a tenth of what current panels cost and at a rate of several hundred feet per minute. Nanosolar's first commercial cells rolled off the presses this year. Cost has always been one of Solar's biggest problems. Traditional solar cells require silicon, and silicon is an expensive commodity. That means even the cheapest solar panels cost about $3 per watt of energy they go on to produce. To compete with coal, that figure has to shrink to just $1 per watt. Nanosolar's cells use no silicon, and the company's manufacturing process allows it to create cells that are as efficient as most commercial cells for as little as 30 cents a watt. "It really is quite a big deal in terms of altering the way we think about solar and in inherently altering the economics of solar," says Dan Kammen, founding director of the Renewable and Appropriate Energy Laboratory at the University of California at Berkeley.

Bribes offered to scientists
2007-02-03, *Sydney Morning Herald* (Australia's leading newspaper)
http://www.smh.com.au/news/environment/bribes-offered-to-scientists/2007/02/0...

Scientists and economists have been offered $10,000 each by a lobby group funded by one of the world's largest oil companies to undermine the UN climate change report. Letters sent by the American Enterprise Institute, an ExxonMobil-funded think tank with close links to the Bush Administration, offered the payments for articles that emphasize the shortcomings of the report. Travel expenses and additional payments were also offered. The institute has received more than $1.6 million from ExxonMobil - which yesterday announced a $50 billion annual profit, the biggest ever by a US company - and more than 20 of its staff have worked as consultants to the Bush Administration. A former head of ExxonMobil, Lee Raymond, is the vice-chairman of the institute's board of trustees.

Note: Why wasn't this important story covered by any major media in the U.S.?

Kids Build Soybean-Fueled Car

2006-02-17, CBS News

http://www.cbsnews.com/stories/2006/02/17/eveningnews/main1329941.shtml

The star at last week's Philadelphia Auto Show wasn't a sports car or an economy car. It was a sports-economy car — one that combines performance and practicality under one hood. But as CBS News correspondent Steve Hartman reports in this week's Assignment America, the car that buyers have been waiting decades [for] comes from an unexpected source and runs on soybean bio-diesel fuel to boot. **A car that can go from zero to 60 in four seconds and get more than 50 miles to the gallon would be enough to pique any driver's interest. So who do we have to thank for it? Ford? GM? Toyota? No — just Victor, David, Cheeseborough, Bruce, and Kosi, five kids from the auto shop program at West Philadelphia High School.** The five kids ... built the soybean-fueled car as an after-school project. It took them more than a year — rummaging for parts, configuring wires and learning as they went. As teacher Simon Hauger notes, these kids weren't exactly the cream of the academic crop. "If you give kids that have been stereotyped as not being able to do anything an opportunity to do something great, they'll step up," he says. Stepping up is something the big automakers have yet to do. They're still in the early stages of marketing hybrid cars while playing catch-up to the Bad News Bears of auto shop. "We made this work," says Hauger. "We're not geniuses. So why aren't they doing it?" Kosi thinks he knows why. The answer, he says, is the big oil companies.

Note: So why isn't this remarkable engine design breakthrough making front page headlines in all major media? Why aren't the many other major energy breakthroughs that have been reported given the headlines they deserve? Could it be that those who are reaping huge profits from oil sales have much more political and media influence than you might imagine?

Eco-car more efficient than light bulb

2005-07-05, CNN News

http://www.cnn.com/2005/TECH/07/04/eco.car

An eco-car that can travel the world using a fraction of the electricity it takes to power a light bulb has been unveiled by its British creators. The hydrogen-powered Ech2o needs just 25 Watts -- the equivalent of less than two gallons of petrol -- to complete the 25,000-mile global trip, while emitting nothing more hazardous than water. But with a top speed of 30mph, the journey would take more than a month to complete. **Ech2o, built by British gas firm BOC, will bid to smash the world fuel efficiency record of over 10,000 miles per gallon** at the Shell Eco Marathon. The record is currently ... 5,385 km/per liter [12,900 mpg!]. John Carolin, BOC global director sustainable energy: "It sounds unbelievable how little power is used to keep the BOC Ech2o moving, but it demonstrates the impact of careful design and is a valuable lesson for car makers in the future.

Note: If these small test cars get over 10,000 miles per gallon, why aren't new cars getting at least 100 mpg?

Refiners Maintain a Firm but Legal Grip on Supplies

2005-06-18, *Los Angeles Times*

http://www.latimes.com/business/la-fi-calgas18jun18,0,7589520.story

California refiners are simply cashing in on a system that allows a handful of players to keep prices high by carefully controlling supplies. The result is a kind of miracle market in which profits abound, outsiders can't compete and a dwindling cadre of gas station operators has little choice but go along. Refiners "not only control how much supply is in the marketplace, they control who gets it and at what price," said Dennis DeCota, executive director of the California Service Station and Automotive Repair Assn. **The recent history of California's fuel industry is a textbook case of how a once-competitive business can become skewed to the advantage of a few, all with the federal government's blessing.** Refiners acknowledge their California businesses have become the most profitable in the nation. The rest of the country isn't far behind. Characteristics once unique to California — specialty fuels, a refinery shortage, the growing dominance of a few companies — have begun to plague other gasoline markets.

Enron: The Smartest Guys in the Room

2005-04-22, *PBS*

http://www.pbs.org/independentlens/enron/film.html

ENRON: The Smartest Guys in the Room [is] the inside story of one of history's greatest business scandals, in which **top executives of America's seventh largest company walked away with over one billion dollars while investors and employees lost everything.** Based on the best-selling book ... this tale of greed, hubris and betrayal reveals the outrageous personal excesses of the Enron hierarchy and the moral vacuum that led CEO Ken Lay - along with other players including accounting firm Arthur Andersen, Chief Operating Officer Jeffrey Skilling and Chief Financial Officer Andy Fastow - to manipulate securities trading, bluff the balance sheets and deceive investors. By 2000, the company has grown into the largest natural gas merchant in North America, eventually branching out into trading other commodities. Jeff Skilling is named CEO, and the company stock skyrockets. Meanwhile, Skilling's "black box" accounting results in declared earnings of 53 million dollars for a collapsing deal that doesn't profit a cent. And **Enron's West Coast power desk has its most profitable month ever as California citizens become casualties of Enron's scheme to artificially increase demand for electricity, resulting in rolling blackouts and two deaths.** When Enron's sleight of hand accounting and unethical trading eventually meet the realities of balance sheets that don't balance and products that don't exist, unwitting employees who have anchored their financial futures to the Enron ship watch in horror as water rushes in overhead.

Note: Enron was American's seventh-largest public company and controlled 25 percent of the nation's energy before it failed in 2002. Its stock plummeted from $90 a share to 9 cents a share in a matter of months after fraud was uncovered.

Toyota smashes fuel economy record

2002-10-20, *London Times*

http://www.timesonline.co.uk/article/0,,588-451038,00.html

Tucked away on the Toyota stand you will find a cheeky little coupé that looks sporty but whose *raison d'ętre* is fuel economy, the lowest exhaust emissions and ease of recycling. **The ES3 — the initials stand for Eco Spirit — achieves 104mpg in the official European fuel consumption tests, a record for a four-seat car.** Some months ago I drove this prototype and not only is it even more economical than the special "3 liter" (three litres of fuel for every 100km travelled, or 94mpg) versions of the Audi A2 and VW Lupo that sell in Germany, but the Toyota is livelier and more responsive and would be very acceptable as an everyday car. The ES3 has a 1.4 litre turbocharged diesel engine and CVT (continuously variable transmission).

Note: So what happened to this amazing car? Why haven't we heard anything about it since the article was published in 2002?

Listen, Detroit: You'll Get a Charge Out of This

1999-03-01, *Time* Magazine

http://www.time.com/time/reports/environment/heroes/heroesgallery/0,2967,ovsh...

Troy, Mich., in the belly of the automobile industry, is an odd place to spark a revolution against the internal-combustion engine. But, then, Stanford Ovshinsky is no ordinary gearhead. Although he never went to college, he founded a new field of physics based on the superconductivity of certain alloys. The company he formed in 1960, Energy Conversion Devices, makes the photovoltaic cells used on the Mir space station to generate electricity from sunlight. In the '80s the Japanese licensed his patents to produce digital video discs. But what really revs him up these days is a car battery. How dull is that? Not at all, if it can "change the world," as he claims with a subversive glint in his eye. When Ovshinsky talked of scaling up his battery to run a car, he was ridiculed. **"The auto companies said it wouldn't work," he recalls. "Then, after one car got 200 miles on a single charge, they said it couldn't be manufactured. Now that we are making them, they say it is too costly. But that is a red herring too.**" Ovshinsky's team of engineers and electrochemists has slashed the cost 40% in two years, they claim. If automakers would commit to buying tens of thousands, Ovshinsky says, the batteries would make electric cars as cheap as gasoline models.

Note: Ovshinsky, who has over 200 patents to his name, was censured for publicizing his amazing battery. GM refused to use his superior battery in the GM's EV-1 when it first came out. The inferior battery they used instead ensured the car would not be successful. Once Ovshinsky's battery became even more effective and looked sure to overtake conventional gasoline as the more effective way to run a car, his company was then sold by GM to Chevron Texaco, who shelved the project entirely.

Method and Apparatus for Tunneling by Melting

1972-09-22, US Patent and Trademark Office

http://patft.uspto.gov/netacgi/nph-Parser?Sect1=PTO1&Sect2=HITOFF&d=PALL&p=1&...

The invention described herein was made in the course of, or under, a contract with the U. S. ATOMIC ENERGY COMMISSION. It relates ... to a method and apparatus for drilling, tunneling and shaft-sinking in rock with particular advantage at hitherto inaccessible depths. The present invention uses the basic apparatus and method disclosed in U.S. Pat. No. 3,357,505 and in Los Alamos Scientific Laboratory of the University of California Report No. LA-3243 (1965) entitled "Rock Melting as a Drilling Technique." In the existing rock melting devices of the prior art, a major difficulty which limited performance was that of delivering a sufficiently large heat flux to the melting face of the drill or penetrator. The development of the heat pipe alleviates this problem in that the use of heat pipes enables the transfer of heat energy from a compact heat source to the extended melting surface of the penetrator at rates high enough to maintain the surface above the melting temperature of the rock. The extrapolation of a mechanism useful for forming large holes in the earth in accordance with **the present invention uses the combination**

of a refractory rock-melting tool, an *in situ* heat source preferably a small nuclear reactor and an exceedingly efficient heat transfer mechanism such as a system of heat pipes to convey heat from the source to the walls of the drilling tool.

Note: This patent shows that government scientists at Los Alamos were using a "small nuclear reactor" to drill underground tunnels. Several of the inventors listed on the patent worked at Los Alamos National Laboratory, including: McInteer, Berthus B.; Mills, Robert L.; Potter, Robert M.; Robinson, Eugene S.; Rowley, John C.; and Smith, Morton C.

Lockheed Martin Receives Patent For 'World Changing' Fusion Reactor

2018-03-28, CBS News (Washington D.C. affiliate)

http://washington.cbslocal.com/2018/03/28/fusion-reactor-power-lockheed-martin/

Lockheed Martin has reportedly been working on a revolutionary new type of reactor that can power anything from cities to aircraft carriers. The Maryland-based defense contractor recently received a patent for the compact fusion reactor (CFR) after filing plans for the device in 2014. According to reports, one generator would be as small as a shipping container but produce the energy to power 80,000 homes or one of the U.S. Navy's Nimitz-class carriers. Lockheed's advanced projects division, Skunk Works, has reportedly been working on the futuristic power source since 2014 and claimed at the time that a CFR could be ready for production by 2019. "I started looking at all the ideas that had been published. I basically took those ideas and melded them into something new by taking the problems in one and trying to replace them with the benefits of others," Dr. Thomas McGuire of Skunk Works said during a 2014 interview. "The nice thing about a fusion reaction is that if somehow it would go out of control, it would just stop itself automatically," William & Mary's Saskia Mordijck told Phys.org in 2012. "If a fission reaction goes out of control, it can really go out of control. You can't stop it and it actually might go into a nuclear meltdown." Lockheed advertises its quest to develop fusion power on its website, calling the technology "a cleaner, safer source of energy" that could be used to power communities or even travel to Mars.

Note: A 2004 *New York Times* article stated that Lockheed Martin runs a "breathtakingly big part" of the US. This company's "Skunk Works" was kept very secret until 2014, when reporters were given a glossy brochure featuring a "10-point "Skunk Works 2015" agenda".

The US government keeps spectacularly underestimating solar energy installation

2017-10-19, *Quartz*

https://qz.com/1103874/the-us-government-underestimated-solar-energy-installa...

Every two years, the US Energy Information Administration (EIA), America's official source for energy statistics, issues 10-year projections about how much solar, wind and conventional energy the future holds for the US. Every two years, since the mid-1990s, the EIA's projections turn out to be wrong. Last year, they proved spectacularly wrong. The Natural Resources Defense Council, an environmental advocacy group, and Statista recently teamed up to analyze the EIA's predictions for energy usage and production. They found that the EIA's 10-year estimates between 2006 to 2016 systematically understated the share of wind, solar and gas. Solar capacity, in particular, was a whopping 4,813% [or 48 times] more in 2016 than the EIA had predicted in 2006 it would be. The EIA regularly underestimates the growth in renewables but *over*estimates US fossil-fuel consumption. These estimates matter because they form the basis for actions by the Environmental Protection Agency and other federal agencies. The agency's "projections bear little resemblance to market realities" because they ignore publicly available evidence, argues the clean-energy non-profit Advanced Energy Economy. Michael Grunwald at Politico reports the EIA seems to base its projections on the assumption that renewable energy costs won't fall much, when in fact they keep plunging.

Tiny turbine that fits on your DESK runs on carbon dioxide - and it can produce enough energy to power a small town

2016-04-12, *Daily Mail* (One of the UK's most popular newspapers)

http://www.dailymail.co.uk/sciencetech/article-3535461/Tiny-turbine-fits-DESK...

Engineers have developed a turbine which has the potential to power a small town all the while being no bigger than your office desk. **Designed by GE Global Research, the turbine could power 10,000 homes and according to researchers, could help to solve some of the world's growing energy challenges. But rather than steam, which is typically used to set turbines in motion, the new turbine uses carbon dioxide**. 'This compact machine will allow us to do amazing things,' said Doug Hofer, lead engineer on the project. According to *MIT Tech Review*, the turbine is driven by 'supercritical carbon dioxide', which is kept under high pressure at temperatures of 700°C. Under these conditions, the carbon dioxide enters a physical state between a gas and a liquid, enabling the turbine to harness its energy for super-efficient power generation - with the turbines transferring 50 per cent of the heat into electricity. It could help energy firms take waste gas and repurpose it for efficient and cleaner energy production. Waste heat produced from other power generation methods, such as solar or nuclear, could be used to melt salts, with the molten salts used to the carbon dioxide gas to a super-critical liquid - which may be much quicker than heating water for steam. Currently, the design of the turbine would enable up to 10,000 kilowatts of energy to be produced, but the turbines could be scaled up to generate 500 megawatts, enough to power a city.

13-Year-Old Creates Energy Harvesting Device

2016-02-18, KTVN News (A leading news channel of Reno, Nevada)

http://www.ktvn.com/story/31260413/13-year-old-creates-energy-harvesting-device

Do you think a 13-year-old could change the world? Max Loughan could be the one to do it. When we interviewed him, Max was wearing his lab coat ... in his parent's old boiler room, which has been converted into a lab. He ponders the future often. "The future that I imagine is the future, frankly we all imagine." He wants to make the world a better place, and to do that, Max believes you need one single thing: "If you got energy, you have power, you have everything." So to solve this problem, a few months ago, Max took the matter into his own hands. He created an electromagnetic harvester out of a coffee can, some wire, two coils, and a spoon. "This cost me 14 bucks," Max said. The harvester conducts radio waves, thermal, and static energy, and turns it into electricity. "This wire takes energy from the air." And the inside the coffee can, "We turn it from AC to DC." **We took the device outside, and wrapped Max's twin brother, Jack, in a string of L.E.D. lights. Max connects the lights to the harvester, and sure enough, they turned on.**

His device clearly works. A $14.00 invention was able to do that. So imagine this same harvester on a scale 20 times larger. "As cheesy as this sounds, from day one, on this planet that I knew I was put here for a reason," said Max. "And that reason is to invent, to bring the future."

Note: Don't miss this video the most amazing 13-year-old who just may have solved the energy problems of our world!!! (https://www.facebook.com/richard.greene.399/videos/10153963366212961/) of

The sports car that runs on saltwater

2014-09-01, *Daily Mail* (U.K.)

http://www.dailymail.co.uk/sciencetech/article-2739768/The-sports-car-runs-SA...

Sports cars may not have the best reputation for being environmentally-friendly, but this sleek machine has been designed to reach 217.5 mph (350 km/h) – using nothing but saltwater. **Its radical drive system allows the 5,070lbs (2,300kg) Quant e-Sportlimousine to reach 0-60 mph (100 km/h) in 2.8 seconds, making it as fast as the McLaren P1. After making its debut at the 2014 Geneva Motor Show in March, the saltwater technology has now been certified for use on European roads.** The 920 horsepower (680 kW) Quant e-Sportlimousine uses something known as an electrolyte flow cell power system to power four electric motors within the car. It works in a similar way to a hydrogen fuel cell, however, the liquid used for storing energy is saltwater. The liquid passes through a membrane in between the two tanks, creating an electric charge. This electricity is then stored and distributed by super capacitors. The car carries the water in two 200-litre tanks, which in one sitting will allow drivers to travel up to 373 miles (600km). NanoFlowcell AG, a Lichtenstein-based company behind the drive, is now planning to test the car on public roads in Germany and elsewhere in Europe as the company prepares for series production. It claims the technology offers five times the energy capacity of lithium-ion batteries of the same weight. 'We've got major plans, and not just within the automobile industry,' says NanoFlowcell AG Chairman of the Board Professor Jens-Peter Ellermann. 'The potential of the NanoFlowcell is much greater, especially in terms of domestic energy supplies as well as in maritime, rail and aviation technology.'

Note: . Why isn't this car and its unique technology getting more press?

Tesla's Model S receives top rating from Consumer Reports

2013-05-09, *Washington Post*/Bloomberg

http://www.washingtonpost.com/business/tesla-electric-model-s-sedan-grabs-con...

Tesla Motors Inc.'s electric Model S, *Motor Trend*'s 2013 "Car of the Year," received the highest rating from *Consumer Reports* in an evaluation of the luxury sedan that led first-quarter North American plug-in car sales. The Model S from Palo Alto, California-based Tesla scored 99 out of 100 points, the non-profit magazine said in an e-mailed statement. **The $89,650 car bought by *Consumer Reports* "performed better, or just as well overall" as any vehicle it's ever tested, the ... magazine said. "It accelerates, handles and brakes like a sports car, it has the ride and quietness of a luxury car and is far more energy efficient than the best hybrid cars,"** said Jake Fisher, Consumer Reports' director of automotive testing. No rechargeable car has won a score as high as the Model S. The magazine last gave a vehicle 99 points in 2007, when Toyota Motor Corp.'s Lexus LS460L ranked that high. Model S shortcomings include limited range, long charge times and "coupe-like styling that impairs rear visibility and impedes access," Consumer Reports said. Along with reliability that isn't yet determined, Tesla still has a limited service network, the magazine said. The test vehicle had an 85-kilowatt/hour lithium-ion battery pack and averaged about 200 miles (322 kilometers) per charge in real-world driving, the magazine said. The Tesla "is easily the most practical electric car that has been tested to date," Consumer Reports said.

Note: After undeniable suppression of the electric car by car manufacturers (Article – Who Killed the Electric Car) , independent upstart Tesla Motors has done it! Expect to see more breakthroughs from this great new company.

Toyota Rav4 EV review: electrifying

2013-03-25, *San Francisco Chronicle* (SF's leading newspaper)

http://www.sfchronicle.com/business/article/Toyota-Rav4-EV-review-electrifyin...

In the new crop of electric cars, the Rav4 may be the best you've never heard of. It comes from one of the world's largest automakers and sports a drivetrain built by Tesla Motors, rock star of the plug-in world. And yet, outside the circle of electric enthusiasts, **few drivers know it exists. You can buy it only in California. Toyota doesn't advertise it on TV. So far, the company has committed to building just 2,600. Critics, including some people who love the Rav4 EV, say Toyota made it only to comply with California regulations that force automakers to sell zero-pollution cars.** "Everyone agrees it's a wonderful car," said Felix Kramer, founder of CalCars, a plug-in vehicle advocacy group. "Too bad there's not enough." That suspicion comes from experience. Toyota made an electric version of the Rav4 once before, building 1,484 of the small SUVs between 1997 and 2003. Then the company killed the program, after California changed its zero-emission vehicle (ZEV) rules. The new Rav4 EV ... boasts ferocious acceleration, plenty of power and a low center of gravity thanks to the big battery pack hidden in the floor. It's not a luxury car, but the interior is comfortable and plush, tricked out with a touch-screen and heated seats. Those so inclined can take the Rav4 EV from a standstill to 60 mph in 7 seconds. The car gets a solid 125 miles on a fully charged battery pack, and an easy-to-read number on the dash constantly reminds you how many miles you have left.

Note: Once again a major car manufacturer produces a great electric vehicle only to suppress it.

Are we wildly underestimating solar and wind power?

2012-06-19, *Washington Post* blog

http://www.washingtonpost.com/blogs/ezra-klein/wp/2012/06/19/are-we-wildly-un...

Right now, renewable energy sources like solar and wind still provide just a small fraction of the world's electricity. But they're growing fast. Solar is growing exponentially. Across the globe, 55 terawatt-hours of solar power had been installed by the end of 2011. That may not seem like much in itself — the United States by itself, after all, needed about one hundred times that much power in 2011. But **solar has been growing at a stunning rate, as panels keep getting dramatically cheaper. If these exponential growth rates [continue] solar could provide nearly 10 percent of the world's electricity by 2018.** Official agencies keep underestimating the growth rate of renewables. The International Energy Agency is forecasting that solar will catch on much more slowly — providing a mere 4.5 percent of the world's electricity by 2035. But [t]he IEA has almost always underestimated how quickly wind and solar can grow. Forecasters have consistently been too pessimistic. For instance, back in 2000, the IEA's World Energy Outlook predicted that non-hydro sources of renewable energy would make up 3 percent of global energy by the year 2020. The world reached that point in 2008, well ahead of schedule. **Using only current technology, renewables could technically provide the vast bulk of U.S. electricity by mid-century.**

Note: The media has consistently underplayed the promising potential for alternative energy sources. The fact that the above is a blog and not a regular article in the *Post* is yet another example of this.

The extra dollars you're paying at the pump are going to Wall Street speculators

2012-02-28, *Chicago Tribune*

http://www.chicagotribune.com/sns-201202280930--tms--amvoicesctnav-a20120228f...

The current surge in gas prices has almost nothing to do with energy policy. It doesn't even have much to do with global supply and demand. It has most to do with America's continuing failure to adequately regulate Wall Street. Oil supplies aren't being squeezed. Over 80 percent of America's energy needs are now being satisfied by domestic supplies. In fact, we're starting to become an energy exporter. Demand for oil isn't rising. Oil demand in the U.S. is down compared to last year at this time. The American economy is showing only the faintest signs of recovery. Meanwhile, global demand is still moderate. Europe's debt crisis hasn't gone away. China's growth continues to slow. But Wall Street is betting on higher oil prices. Hedge-fund managers and traders assume that mounting tensions in the Middle East will hobble supplies later this year. Wall Street speculators also assume global demand for oil will rise in the coming year. These are just expectations, not today's realities. But they're pushing up oil prices just the same, because Wall Street firms and other big financial players now dominate oil trading. Where there's money to be made, Wall Street will find a way of making it. And when it comes to oil, **so much money is at stake that gigantic sums can be made if the bets pay off. Speculators figure they can hedge against bad bets. Financial speculators historically accounted for about 30 percent of oil**

contracts, producers and end users for about 70 percent. But today speculators account for 64 percent of all contracts.

Note: This article was written by Robert Reich, former U.S. Secretary of Labor, professor of public policy at the University of California at Berkeley and the author of *Aftershock: The Next Economy and America's Future*. He blogs at www.robertreich.org

Method for Enhancement of Surface Plasmon Polaritons to Initiate & Sustain LENR (Cold Fusion)

2012-01-16, NASA Website's Technology Gateway

http://technologygateway.nasa.gov/media/CC/lenr/lenr.html

[Video transcript] Narrator: While the world is drastically dependent on fossil fuel, researchers at NASA Langley Research Center are working on another way of producing energy-efficient nuclear power. Senior Research Scientist Dr Zawodny: This other form of nuclear power releases energy by adding neutrons. Eventually [the nuclei] gain a sufficient number of neutrons that they spontaneously decay into something of the same mass but a different element. **It has the demonstrated ability to produce excess amounts of energy, cleanly, without hazardous ionizing radiation, without producing nasty waste. Narrator: This clean form of energy is ... able to support everything from transportation systems to infrastructure. Dr. Zawodny: The easiest implementation of this would be for the home.** It would be ... dual use. It would [produce] heat; and you'd derive electricity from it to run your electronics, power the house, power the building, power the light industry. And then the waste heat would be used for environmental control [i.e. heating, air conditioning, etc.] and warm water." Narrator: NASA's method for enhancement of surface plasmon polaritons to initiate and sustain LENR in Metal Hydride Systems, a clean nuclear energy for your power-operated technology.

Note: LENR stands for Low Energy Nuclear Reactions, aka cold fusion. So NASA is now acknowledging cold fusion is real! And their research provides major hope for the future.

Solar Energy: The Quest for Cheap

2011-10-13, Bloomberg Businessweek

http://www.businessweek.com/technology/solar-energy-the-quest-for-cheap-10132...

The big number is 50. **When companies can produce solar photovoltaic modules for less than 50˘ per watt, solar energy will be able to compete directly with coal. Right now, the cheapest solar cells are being produced for as little as 70˘ per watt.** They are selling for about $1.26 per watt, with prices expected to drop to $1.17 next year. **Most anticipate they will hit 50˘ per watt within four or five years.** As prices fall, demand is growing. Total solar installations in the second quarter [of 2011] grew by 69 percent over the same period in 2010. The number of Americans working in the solar industry more than doubled, to 100,000, from 2009 to 2011. That's considerably more than the 80,600 coal miners working in the U.S. Behind the price drops are cheaper manufacturing costs, lower costs for such crucial raw materials as silicon, and rapidly improving technology. Dozens of startups in the U.S. have potentially transformative ideas. The question is which can come out on top. The wide variety of companies developing competing technologies to capture and distribute solar power underscores the market's immaturity. Currently, researchers are experimenting with materials ranging from silicon to gallium arsenide to cadmium telluride, basing cost projections on disparate technologies that create solar cells. The goal is to build one that competes without government subsidies.

Man makes drinking water out of thin air

2011-08-01, KSDK.com/CNN

http://www.ksdk.com/news/article/269771/28/Man-makes-drinking-water-out-of-th...

Water out of air? A Texas man has invented a machine that does just that. The drought doesn't worry [inventor Terry LeBleu] because he has invented and patented a new machine. It's called the "Drought Master" and makes drinkable water out of air. "It pulls the air through it, pulls out the moisture, and exhausts the air," LeBleu says. **Depending on humidity, the machine can make between five to seven gallons of pure water in one day. All you have to do is plug it in, and one gallon costs only 4 cents in electrical charges.** An independent lab took samples of LeBleu's water and found it had no bacteria and is free of metals. Lab techs say it's similar to distilled water. Willie Nelson owns 50 of these machines, including an indoor version. Even Texas Governor Rick Perry owns one. But LeBleu wants his invention to benefit local farmers and ranchers. The machine is quieter than a refrigerator, and you only have to wash the filter every few years. Building one takes only two hours. The oldest model made is still up and running. It's been functioning for a decade.

Who Killed the Electric Car?

***Popular Science* magazine, July 10, 2006**
Posted: 2010-05-24 20:28:25

http://www.popsci.com/cars/article/2006-07/who-killed-electr...

Chris Paine´s documentary film "Who Killed the Electric Car?" argues convincingly that there was indeed a market for the cars — and a devoted one, ... but that **GM [General Motors] squashed the EV1 because, quite simply, it threatened the livelihood of the entire automotive industry. The car used no gasoline, no oil and no mufflers, and it required only sporadic brake maintenance. Each of these components represents billions of dollars in profits for the industry.** GM, the oil companies and various government agencies argued that the car wasn´t practical, didn´t have enough range for consumers and was less promising than the apparently imminent hydrogen technology. The reality was exactly the opposite, Paine´s film suggests — the viability of hydrogen as an automotive fuel source alone is in fact almost comically optimistic. The whisper-quiet EV1 was designed by [an] aviation pioneer, Paul MacCready of AeroVironment. In the 1970s, MacCready built the only successful human-powered aircraft, the Gossamer Condor and the Gossamer Albatross. His solar-powered electric car Sunraycer, built for GM, won the 1987 World Solar Challenge Race in Australia. His corporate mantra is "do more with less" — that is, focus on creating vehicles that require less energy to operate, not on finding ways to pump more power into inefficient systems. His team´s battery-powered EV1 was a triumph of engineering and a joy to operate.

Safe nuclear does exist, and China is leading the way with thorium

2011-03-20, *The Telegraph* (One of the UK's leading newspapers)

http://www.telegraph.co.uk/finance/comment/ambroseevans_pritchard/8393984/Saf...

A few weeks before the tsunami struck Fukushima's uranium reactors and shattered public faith in nuclear power, China revealed that it was launching a rival technology to build a safer, cleaner, and ultimately cheaper network of reactors based on thorium. China's Academy of Sciences said it had chosen a "thorium-based molten salt reactor system". The liquid fuel idea was pioneered by US physicists at Oak Ridge National Lab in the 1960s. Chinese scientists claim that hazardous waste will be a thousand times less than with uranium. The system is inherently less prone to disaster.**"The reactor has an amazing safety feature," said Kirk Sorensen, a former NASA engineer at Teledyne Brown and a thorium expert. "If it begins to overheat, a little plug melts and the salts drain into a pan. There is no need for computers, or the sort of electrical pumps that were crippled by the tsunami. The reactor saves itself,"** he said. US physicists in the late 1940s explored thorium fuel for power. It has a higher neutron yield than uranium, a better fission rating, longer fuel cycles, and does not require the extra cost of isotope separation. The plans were shelved because thorium does not produce plutonium for bombs. As a happy bonus, it can burn up plutonium and toxic waste from old reactors, reducing radio-toxicity and acting as an eco-cleaner.

Note: For a 30-minute documentary on the powerful potential of thorium as an energy source, click here. For many reports from reliable sources on promising new energy technologies, click here.

Fusion's Ups and Downs

2010-03-23, MSNBC

http://cosmiclog.msnbc.msn.com/archive/2010/03/23/2237165.aspx

This week, scientists gathered at the American Chemical Society's spring meeting in San Francisco to turn the spotlight on a highly unorthodox path: the effect known as cold fusion. This year's session featured nearly 50 presentations - including reports on batteries and bacteria that appear to exhibit the cold-fusion effect. Back in 1989, cold fusion was heralded as a simple, inexpensive way to get a power-generating fusion reaction on a desktop. But when the experimental results couldn't be reproduced, the researchers were driven into obscurity [and] the term "cold fusion" became synonymous with quackery. Chemists, however, have kept up their interest in the effect. Rick Nebel [has headed] up a handful of researchers following the less-traveled path to fusion at EMC2 Fusion Development Corp. EMC2 recently created a buzz in the fusion underground by reporting on its Web site that it successfully completed a series of experiments to "validate and extend" earlier results. The company is now using a $7.9 million contract from the U.S. Navy to build a bigger test machine. **Nebel and his colleagues are now seeking contributions to fund the development of what they say would be a 100-megawatt fusion plant - a "Phase 3" effort projected to cost $200 million and take four years. "Successful Phase 3 marks the end of fossil fuels," the Web site proclaims.**

The Bloom Box: An Energy Breakthrough?

2010-02-18, CBS 60 Minutes

http://www.cbsnews.com/stories/2010/02/18/60minutes/main6221135.shtml

In the world of energy, the Holy Grail is a power source that's inexpensive and clean, with no emissions. Over 100 start-ups in Silicon Valley are working on it. One of them, **Bloom Energy, is about to make public its invention: a little power-plant-in-a-box they want to put literally in your backyard. You'll generate your own electricity with the box and it'll be wireless. The idea is to one day replace the big power plants and transmission line grid**. K.R. Sridhar ... says he knows it works because he originally invented a similar device for NASA. He really is a rocket scientist. He invented a new kind of fuel cell, which is like a very skinny battery that always runs. Sridhar feeds oxygen to it on one side, and fuel on the other. The two combine within the cell to create a chemical reaction that produces electricity. There's no need for burning or combustion, and no need for power lines from an outside source. "It's cheaper than the grid, it's cleaner than the grid." **Twenty large, well-known companies have quietly bought and are testing Bloom boxes in California. The first customer was Google.** Four units have been powering a Google datacenter for 18 months. They use natural gas, but half as much as would be required for a traditional power plant. John Donahoe, eBay's CEO, says its five boxes were installed nine months ago and have already saved the company more than $100,000 in electricity costs. eBay's boxes run on bio-gas made from landfill waste, so they're carbon neutral. "In five to ten years, we would like to be in every home." [Sridhar] said a unit should cost an average person less than $3,000.

Note: To watch the fascinating 60 Minutes video clip of this amazing invention, click here. For other CBS videos clips on the Bloom Box, click here. For astounding news on other new energy sources and inventions, click here and here.

Solar Energy, All Night Long

2008-07-31, *Forbes* Magazine

http://www.forbes.com/2008/07/30/nocera-solar-power-biz-energy-cz_jf_0731sola...

MIT professor Daniel G. Nocera has long been jealous of plants. He desperately wanted to do what they do--split water into hydrogen and oxygen and use the products to do work. That, he figures, is the only way we humans can solve our energy problems; **enough energy pours down from the sun in one hour to power the planet's energy needs for a year. Nocera's discovery [is] a cheap and easy way to store energy that he thinks will be used to change solar power into a mainstream energy source.** Plants catch light and turn it into an electric current, then use that energy to excite catalysts that split water into hydrogen and oxygen during what is called photosynthesis' light cycle. The energy is then used during the dark cycle to allow the plant to build sugars used for growth and energy storage. Nocera and Matthew Kanan, a postdoctoral fellow in Nocera's lab, focused on the water-splitting part of photosynthesis. They found cheap and simple catalysts that did a remarkably good job. They dissolved cobalt and phosphate in water and then zapped it with electricity through an electrode. The cobalt and phosphate form a thin-film catalyst around the electrode that then use electrons from the electrode to split the oxygen from water. The oxygen bubbles to the surface, leaving a proton behind. A few inches away, another catalyst, platinum, helps that bare proton become hydrogen. The hydrogen and oxygen, separated and on-hand, can be used to power a fuel cell whenever energy is needed.

Note: This amazing breakthrough resulted in a $4 million government grant for further development.

Cars that make hybrids look like gas guzzlers

2007-03-04, *San Francisco Chronicle* (San Francisco's leading newspaper)

http://www.sfgate.com/cgi-bin/article.cgi?f=/c/a/2007/03/04/ING44OD4AS1.DTL

Toyota Prius owners tend to be a proud lot since they drive the fuel-efficient hybrid gas-electric car that's ... one of the hottest-selling vehicles in America. A few, however, felt that good was not good enough. They've made "improvements" even though the modifications voided parts of their warranties. Why? Five words: one hundred miles per gallon. "We took the hybrid car to its logical conclusion," [Felix] Kramer says, by adding more batteries and the ability to recharge by plugging into a regular electrical socket at night. Compared with the Prius' fuel efficiency of 50 mpg, plug-in hybrids use half as much gasoline by running more on cleaner, cheaper, domestic electricity. **These trendsetters monkeyed with the car ... to make a point: If they could make a plug-in hybrid, the major car companies could, too.** Kramer ... and a cadre of volunteers formed the California Cars Initiative (online at calcars.org). They added inexpensive lead-acid batteries ... giving the car over 100 mpg in local driving and 50 to 80 mpg on the highway. The cost of conversion is about $5,000 for a do-it-yourselfer. Several small companies like EnergyCS ... started doing small numbers of conversions for fleets and government agencies using longer-lasting, more energy-dense lithium-ion batteries. Kramer hired EnergyCS to convert his Prius and reported on a typical day of driving. Compared with driving his Prius before the conversion, he ... spewed out two-thirds less greenhouse gases at a total cost of $1.76 for electricity and gasoline, instead of the $3.17 it would have required on gasoline alone. People want plug-in hybrids but can't get them. Dealers don't sell them yet, and the few conversion services cater to fleets.

Note: For a video and educational package to guide those who want to build a 100 mpg car, see www.eaa-phev.org. For why the car companies with their massive budgets haven't developed cars like this, click here.

Coming in out of the cold: Cold fusion, for real

2005-06-06, *Christian Science Monitor*

http://www.christiansciencemonitor.com/2005/0606/p25s01-stss.html

A very reputable, very careful group of scientists at the University of Los Angeles ... has initiated a fusion reaction using a laboratory device that's not much bigger than a breadbox, and works at roughly room temperature. This time, it looks like the real thing. The whole trick with fusion is you've got to get protons close enough together for the strong force to overcome their electrical repulsion and merge them together into a nucleus. Instead of using high temperatures and incredible densities to ram protons together, the scientists at UCLA cleverly used the structure of an unusual crystal. Crystals are fascinating things; the atoms inside are all lined up in a tightly ordered lattice, which creates the beautiful structure we associate with crystals. Stressing the bonds between the atoms of some crystals causes electrons to build up on one side, creating a charge difference over the body of the crystal. Instead of using intense heat or pressure to get nuclei close enough together to fuse, this new experiment used a very powerful electric field to slam atoms together. This experiment has been repeated successfully and other scientists have reviewed the results. For the time being, don't expect fusion to become a readily available energy option. The current cold fusion apparatus still takes much more energy to start up than you get back out. But it really may not be long until we have the first nuclear fusion-powered devices in common use.

Note: If the above link fails, click here. Why wasn't this widely reported? For a possible answer, click here.

Fans of GM Electric Car Fight the Crusher

2005-03-10, *Washington Post*

http://www.washingtonpost.com/wp-dyn/articles/A21991-2005Mar9.html

What's at stake, they say, is no less than the future of automotive technology, a practical solution for driving fast and fun with no direct pollution whatsoever. **GM agrees that the car in question, called the EV1, was a rousing feat of engineering that could go from zero to 60 miles per hour in under eight seconds with no harmful emissions.** The market just wasn't big enough, the company says, for a car that traveled 140 miles or less on a charge before you had to plug it in like a toaster. Some 800 drivers once leased EV1s, mostly in California. After the last lease ran out in August, GM reclaimed every one of the cars, donating a few to universities and car museums but crushing many of the rest. Enthusiasts discovered a stash of about 77 surviving EV1s behind a GM training center in Burbank and last month decided to take a stand. Mobilized through Internet sites and word of mouth, nearly 100 people pledged $24,000 each for a chance to buy the cars from GM. On Feb. 16 the group set up a street-side outpost of folding chairs that they have staffed ever since in rotating shifts, through long nights and torrential rains, trying to draw attention to their cause. GM refuses to budge. Toyota is aware of a growing fad among do-it-yourselfers who put a new battery in their Prius so it can be plugged in at home and then travel about 20 miles on electric power alone.

Note: Why would GM simply crush cars for which people are willing to pay $24,000? For a possible answer to this important question, click here. To learn how to convert a Toyota Prius to get 100 mpg, click here.

Air Force pursuing antimatter weapons

2004-10-04, *San Francisco Chronicle* (San Francisco's leading newspaper)

http://www.sfgate.com/cgi-bin/article.cgi?file=/c/a/2004/10/04/MNGM393GPK1.DTL

The U.S. Air Force is quietly spending millions of dollars investigating ways to use a radical power source -- antimatter, the eerie "mirror" of ordinary matter -- in future weapons. The most powerful potential energy source presently thought to be available to humanity, antimatter is a term normally heard in science-fiction films. But antimatter itself isn't fiction. During the Cold War, the Air Force funded numerous scientific studies of the basic physics of antimatter. **Following an initial inquiry from The Chronicle this summer, the Air Force forbade its employees from publicly discussing the antimatter research program. Still, details on the program appear in numerous Air Force documents distributed over the Internet prior to the ban.** It almost defies belief, the amount of explosive force available in a speck of antimatter. One millionth of a gram of positrons contain as much energy as 37.8 kilograms (83 pounds) of TNT. A simple calculation, then, shows that about 50-millionths of a gram could generate a blast equal to the explosion ... in Oklahoma City in 1995. Officials at Eglin Air Force Base initially agreed enthusiastically to try to arrange an interview with ... Kenneth Edwards, director of the "revolutionary munitions" team at the Munitions Directorate at Eglin. "We're all very excited about this technology," spokesman Rex Swenson [said] in late July. But Swenson backed out in August after he was overruled by higher officials in the Air Force and Pentagon. Reached by phone in late September, Edwards repeatedly declined to be interviewed. His superiors gave him "strict instructions not to give any interviews personally. "I'm sorry about that -- this (antimatter) project is sort of my grandchild."

This breakthrough in a type of photosynthesis could provide the world with unlimited energy

2018-09-12, *MarketWatch*

https://www.marketwatch.com/story/this-breakthrough-in-a-type-of-photosynthes...

Imagine a new, potent generation of solar panels capable of producing unlimited amounts of energy, using only sunshine and algae. This could be possible, thanks to a breakthrough made by researchers from the University of Cambridge, documented in a *Nature Energy 2018*article. They were able to split water into its components, oxygen and hydrogen, using what is known as semi-artificial photosynthesis. The procedure has ... never been used to generate large amounts of energy due to expensive and toxic catalysts necessary for the reaction. Photosynthesis [is] the process plants use to convert sunlight into energy. Oxygen is a byproduct of photosynthesis when water absorbed by plants is "split." Most of the oxygen on Earth is here because of this photochemical reaction. Hydrogen ... is also produced this way. Now, by combining algae and man-made components, researchers have been able to bypass both natural inefficiency and the use of toxic reactants. This was achieved by enabling a dormant process in algae that uses a special enzyme (hydrogenase) to reduce water into hydrogen and oxygen. Katarzyna Sokol, a researcher on the project ... explains: "Hydrogenase is an enzyme present in algae that is capable of reducing protons into hydrogen. During evolution, this process has been deactivated because it wasn't necessary for survival, but we successfully managed to bypass the inactivity to [split] water into hydrogen and oxygen."

We’re drowning in trash. These Dutch scientists have a solution.

2018-08-02, *Washington Post*

https://www.washingtonpost.com/news/theworldpost/wp/2018/08/02/biomass/?nored...

The sprawling, gated campus of the Energy Research Center of the Netherlands (ECN) sits on a spit of land about an hour north of Amsterdam. In a nearby control room, engineers ... were working on one of clean energy’s intransigent problems: how to turn waste into electricity without producing more waste. Decades ago, scientists discovered that when heated to extreme temperatures, wood and agricultural leftovers, as well as plastic and textile waste, turn into a gas composed of underlying chemical components. The resulting synthetic gas, or “syngas,” can be harnessed as a power source, generating heat or electricity. But gasified waste has serious shortcomings: it contains tars, which clog engines and disrupt catalysts, breaking machinery, and in turn, lowering efficiency and raising costs. This is what the Dutch technology is designed to fix. The MILENA-OLGA system, as they call it, is a revolutionary carbon-neutral energy plant that turns waste into electricity with little or no harmful byproducts. **The MILENA-OLGA process ... is 11 percent more efficient than most existing energy-from-waste plants and over 50 percent more efficient than incinerators of a comparable scale. The process also emits zero wastewater and produces no particulates or other pollutants**. Just 4 percent of the original material is left over as inert white ash, which can be used to make cement.

Note: A similar technology was developed and implemented over 10 years ago, as detailed in this *Popular Science* article. Why wasn't this amazing invention widely reported and used?

Costa Rica to ban fossil fuels and become world's first decarbonized society

2018-05-10, *The Independent* (One of the UK's leading newspapers)

https://www.independent.co.uk/environment/costa-rica-fossil-fuels-ban-preside...

Costa Rica's new president has announced a plan to ban fossil fuels and become the first fully decarbonized country in the world. Carlos Alvarado, a 38-year-old former journalist, made the announcement ... during his inauguration. "Decarbonization is the great task of our generation and Costa Rica must be one of the first countries in the world to accomplish it, if not the first," Mr. Alvarado said. **Symbolically, the president arrived at the ceremony in San Jose aboard a hydrogen-fueled bus. Last month, Mr. Alvarado said the Central American country would begin to implement a plan to end fossil fuel use in transport by 2021 – the 200th year of Costa Rican independence**. "When we reach 200 years of independent life we will take Costa Rica forward and celebrate ... that we've removed gasoline and diesel from our transportation," he promised during a victory speech. **Costa Rica already generates more than 99 per cent of its electricity using renewable energy sources.** Costa Rica's push towards clean energy faces no large-scale backlash, in part because the country has no significant oil or gas industry. But demand for cars is rising, as is use of other transport systems, and that may prove one of the biggest challenges in meeting the new goal. Transport is today the country's main source of climate changing emissions.

Note: Explore a treasure trove of concise summaries of incredibly inspiring news articles which will inspire you to make a difference.

Power Prices Go Negative in Germany, a Positive for Energy Users

2017-12-25, *New York Times*

https://www.nytimes.com/2017/12/25/business/energy-environment/germany-electr...

Germany has spent $200 billion over the past two decades to promote cleaner sources of electricity. That enormous investment is now having an unexpected impact - consumers are now actually paid to use power on occasion, as was the case over the weekend. Power prices plunged below zero for much of Sunday and the early hours of Christmas Day on ... a large European power trading exchange, the result of low demand, unseasonably warm weather and strong breezes that provided an abundance of wind power on the grid. Such "negative prices" are not the norm in Germany, but they are far from rare, thanks to the country's effort to encourage investment in greener forms of power generation. **Prices for electricity in Germany have dipped below zero ... more than 100 times this year alone.** Several countries in Europe have experienced negative power prices, including Belgium, Britain, France, the Netherlands and Switzerland. But Germany's forays into negative pricing are the most frequent. At times, Germany is able to export its surplus electricity to its neighbors, helping to balance the market. Still, its experiences of negative prices are often longer, and deeper, than they are in other countries. In one recent example, power prices spent 31 hours below zero during the last weekend of October. At one point, they dipped as low as minus €83, or minus $98, per megawatt-hour, a wholesale measure. Anyone who was able to hook up for a large blast of electricity at that time was paid €83 per unit for the trouble.

This Incredible Power Plant Produces Nothing But Electricity And Stone

2017-10-16, *Huffington Post*

http://www.huffingtonpost.co.uk/entry/hellisheidi-power-plant-produces-nothin...

The Hellisheidi geothermal power plant situated on the mid-Atlantic ridge, is the newest and largest geothermal plant in Iceland - a country that heats 90% of its homes using geothermal water. **The plant has now become the first in history to capture carbon dioxide from ambient air, using a system of fans and filters, and then store it in bedrock 700 metres down**. There the gas reacts with basaltic rock and forms solid minerals, creating a permanent storage solution, and turning Hellisheidi into a negative emissions site. The EU-funded project [is] capable of capturing 50 metric tons of CO2 each year. Christoph Gebald, Founder and CEO at Climeworks, said: "The potential of scaling-up our technology in combination with CO2 storage, is enormous. Our plan is to offer carbon removal to individuals, corporates and organizations as a means to reverse their non-avoidable carbon emissions.' It also costs $600 per ton of carbon dioxide, a figure they are hoping to reduce to $100 per ton. Iceland currently runs 100% of it's electricity from renewable sources.

GM is selling a $5,000 electric car in China

2017-08-08, CNN News

http://money.cnn.com/2017/08/07/autos/gm-china-electric-car/?sr=twCNN080717gm...

General Motors will start selling a tiny electric car in China this week that will cost about $5,300 after national and local electric vehicle incentives. For that sort of price, the Baojun E100 is no Cadillac, of course. The two-seat car's wheelbase - the distance from the center of the front wheels to the center of the rear wheels - is just 63 inches. Prices for the car start at RMB 93,900, or about $14,000, before incentives. The E100, which is Baojun's first electric car, is powered by a single 39-horsepower electric motor and has a top speed of 62 miles an hour. The E100 can drive about 96 miles on a fully charged battery. Baojun is a mass-market car brand from General Motors' SAIC-GM-Wuling joint venture in China. It's China's eighth most popular car brand. **More than 5,000 people have already registered to buy the first 200 vehicles**, according to GM. Another 500 vehicles will be made available this week, and buyers will be chosen on a first-come-first-served basis, a GM spokesperson said. Sales will initially be limited to the Guanxi region of southern China, but GM plans to sell the car more widely in China. A GM spokesperson declined to say exactly how many it expects to sell. China is the largest automotive market in the world, and its government is making a big push for electric cars. Already, China accounts for 40% of all electric cars sold worldwide, according to the International Energy Agency.

Scientists flip energy equation with solar leaf that converts CO2 into fuel

2016-08-01, *Christian Science Monitor*

http://www.csmonitor.com/Science/2016/0801/Scientists-flip-energy-equation-wi...

It's often smarter to borrow from nature than reinvent the wheel. That was the approach of researchers at the University of Illinois at Chicago (UIC) to remove carbon dioxide (CO_2) from the atmosphere, and convert it into an efficient, inexpensive fuel. The result: an artificial leaf that turns CO_2 into fuel, "at a cost comparable to a gallon of gasoline" could render fossil fuel obsolete, according to the researchers. The "leaf" is one of a growing number of inventions that mimic photosynthesis to remove excess carbon from the atmosphere, and convert it into new, sustainable forms of energy to power our world. "**The new solar cell is not photovoltaic - it's photosynthetic," said [the study's lead author] Amin Salehi-Khojin. "Instead of producing energy in an unsustainable one-way route from fossil fuels to greenhouse gas, we can now reverse the process and recycle atmospheric carbon into fuel using sunlight**." The concept of reduction reaction - converting CO_2 into a burnable form of carbon - isn't new. But scientists previously relied on silver and other expensive precious metals to break gas into storable energy. UIC researchers took a different approach. When light strikes the "leaf," hydrogen and carbon monoxide bubble from the cathode, while free oxygen and hydrogen ions are released from the anode. Leafs could be spread throughout a solar farm, or used in smaller applications, the researchers said.

Long-lasting flow battery could run for more than a decade with minimum upkeep

2016-06-20, Harvard School of Engineering and Applied Sciences

https://www.seas.harvard.edu/news/2017/02/long-lasting-flow-battery-could-run...

Researchers from the Harvard John A. Paulson School of Engineering and Applied Sciences (SEAS) have developed a new flow battery that stores energy in organic molecules dissolved in neutral pH water. This new chemistry allows for a non-toxic, non-corrosive battery with an exceptionally long lifetime and offers the potential to significantly decrease the costs of production. Flow batteries store energy in liquid solutions in external tanks - the bigger the tanks, the more energy they store. Flow batteries are a promising storage solution for renewable, intermittent energy like wind and solar but today's flow batteries often suffer degraded energy storage capacity after many charge-discharge cycles. The Harvard team was able to engineer a battery that loses only one percent of its capacity per 1000 cycles. "Lithium ion batteries don't even survive 1000 complete charge/discharge cycles," said [researcher Michael] Aziz. "Because we were able to dissolve the electrolytes in neutral water, this is a long-lasting battery that you could put in your basement," said [researcher Roy] Gordon. "If it spilled on the floor, it wouldn't eat the concrete and since the medium is noncorrosive, you can use cheaper materials to build the components of the batteries, like the tanks and pumps." The Department of Energy (DOE) has set a goal of building a battery that can store energy for less than $100 per kilowatt-hour. "If you can get anywhere near this cost target then you change the world," said Aziz.

NASA approves 'impossible' space engine design that apparently violates the laws of physics and could revolutionize space travel

2014-08-04, *The Independent* (One of the UK's leading newspapers)

http://www.independent.co.uk/news/science/nasa-approves-impossible-space-engi...

In a quiet announcement that has sent shockwaves through the scientific world, NASA has cautiously given its seal of approval to a new type of "impossible" engine that could revolutionize space travel. In a paper published by the agency's experimental Eagleworks Laboratories, NASA engineers confirmed that they had produced tiny amounts of thrust from an engine without propellant – an apparent violation of the conservation of momentum; the law of physics that states that every action must have an equal and opposite reaction. **NASA's engineers have tested an engine known as a 'Cannae Drive', a machine [that] uses electricity to generate microwaves, bouncing them around inside a specially designed container that theoretically creates a difference in radiation pressure and so results in directional thrust.** In an ordinary engine the rocket moves forward as fuel is flung backwards - the momentum of the rocket (a measure of both its mass and velocity combined) is 'conserved' because it is moved from the rocket to the fuel. However, with the Cannae Drive there is no fuel - the microwaves aren't expelled from the engine. NASA's scientists tested a version of the drive designed by US scientist Guido Fetta and found that the propellantless engine was able to produce between 30 and 50 micronewtons of thrust – a tiny amount (0.00003 to 0.00005 per cent of the force of an iPhone pressing down when held in the hand) but still a great deal more than nothing.

Power shifts as off-grid options spread worldwide

2012-04-13, *San Francisco Chronicle*/Bloomberg

http://www.sfgate.com/cgi-bin/article.cgi?f=/c/a/2012/04/12/BU031O2FFD.DTL

An electricity revolution [is] sweeping through power markets and threatening traditional utilities' dominance of the world's supply. From the poorest parts of Africa and Asia to the most-developed regions in the United States and Europe, solar units ... and small-scale wind and biomass generators promise to extend access to power to more people than ever before. In the developing world, they're slashing costs in the process. Across India and Africa, startups and mobile phone companies are developing what are called called microgrids, in which stand-alone generators power clusters of homes and businesses in places where electric utilities have never operated. In Europe, cooperatives are building their own generators and selling power back to the national or regional grid. The revolution is just beginning, says Jeremy Rifkin, a professor at the Wharton School of the University of Pennsylvania and author of "The Third Industrial Revolution." Disruptive to the economic status quo, **the transformation opens up huge opportunities to consumers who may find themselves trading power in the future much as they swap information over the Internet today, he says. "This is power to the people." Within a decade, installing photovoltaic panels may be cheaper for many families than buying power from national grids in much of the world**, including the United States, Japan, Brazil and the United Kingdom, according to data from Bloomberg New Energy Finance.

Section Seven

Conclusions

Building a Brighter Future

All the darkness of the world cannot put out the light of a single candle.

The Current Situation

Together, we can and will build a brighter future. Yet in order to give meaningful suggestions to further support this, we first need to speak candidly about what's happening in the world at present. It appears that there are factions within the world's power elite which still desire to exert as much control as possible over the world.

Their primary means for establishing control are through promoting fear, secrecy, and polarization, and through distracting people from their deeper purpose in life. When these factions gain greater power and control, our freedoms and liberties are increasingly taken away, sometimes without our even realizing it.

Fear is used as a powerful tool of control by these factions of the global elite. "The terrorists want to kill us all. Your job, savings, and retirement are no longer secure. You are either for us or against us. We might be attacked at any time." All of these messages push us towards fear. And by encouraging us to focus blame on others such as terrorists and "evil" leaders, the global power brokers push us away from feeling in control of our lives and towards the role of powerless victims, so that we look to them for protection. The more we slip into fear and being victims, the easier it is for us to be manipulated.

Secrecy leads to control through preventing the exposure of hidden agendas, and through breeding distrust, suspicion, and paranoia in the world. In the name of "national security," we have been told ever more frequently that we should not know what is happening behind closed doors in government. Yet we are encouraged to keep vigilant watch over our neighbors, over those who question the government, and over those who look or act different from us, as they could secretly be terrorists in sheep's clothing. Rampant suspicion and secrecy cause us to lose touch with the common humanity we share with all around us.

Polarization leads to "us vs. them" ways of thinking. Divide and conquer is a tactic that has been used successfully by the power elite of the world for centuries. As long as people focus on blaming and attacking "the enemy," they are not likely to look for the deeper causes of their problems. Controlling factions of the power elite work to keep public and media attention focused on issues which divide us into two polarized camps, so that their greatest concerns of power and control go unnoticed. The power brokers then control both sides of the divided public by assuring that corporate ownership of the media and key candidates from all political parties support their deeper power and polarization agendas, which often involve war profiteering.

A final subtle, yet powerful method of gaining control is to distract people from their purpose in life. Consider that television, radio, and movies have been filled with ever more greed, violence, and empty sex. Our educational system has increasingly prioritized passing tests over developing intelligence and creativity. Excessive corporate control of the media has limited what news we receive, and often shaped our decisions and the way we think. Our attention has been subtly diverted from our deeper purpose in life to the more superficial attractions of sex, consumerism, and money.

When we succumb to fear, secrecy, and polarization, and when we lose touch with our sense of purpose in life, elements of the global elite are able to exert ever more control over our lives and world.

Towards a Brighter Future

What can we do about all this? Ultimately, it is the collective fear, secrecy, polarization, and loss of purpose within all of us that has allowed leaders to take power who would subvert democracy, and who would take away our freedoms and liberties. By each one of us making a commitment to work on these issues both inside of ourselves and out in the world – and to inspire others to do the same – we are changing our collective direction and building a solid foundation for a brighter future. Several actions can help us to take firm steps in that direction.

Develop purpose and intentions for your life. To live rich, full lives in today's complex world, it's vitally important to give your life clear direction by exploring what is deepest and most meaningful to you, and by developing a life purpose and life intentions based on this. Then choose to live your intentions and follow your purpose to the best of your ability every day. By choosing to live with clear direction and focus, life becomes deeper and more meaningful. This then weakens the seduction of consumerism and media hype which distract us from our purpose, and allows us to more effectively focus on building a brighter future.

Choose to transform fear into acceptance and love. When we notice ourselves feeling fear, we can trace the roots of that fear, so that we are able to identify our core challenges and deal more consciously with them. We can open to guidance from friends, teachers, and spiritual sources in helping us to move from fear to acceptance and understanding. We can also transform our fears through welcoming the ever-present love of our divine creator, and the love that lies always deep within our own hearts, and in the hearts of those around us. In doing this, we begin to recognize fear as an invitation to growth.

Become aware of when you are playing the role of victim, and choose instead to take personal responsibility for building a brighter future. In blaming others for our problems, we often avoid taking responsibility for how much we are involved in creating these problems through the choices we make. Whenever we catch ourselves playing the role of victim by blaming others (including the power elite) for everything that's wrong in our lives, we can choose to take a look inside ourselves, and to explore and take responsibility for our role in what's happening. By focusing less on blaming others, and more on improving ourselves, every one of us can make a difference both in our lives and in our world.

Avoid secrecy and encourage openness and transparency. An important way we can do this is to work together to inform friends and colleagues about the major cover-ups being hidden from the public. Then in our personal lives, when we notice ourselves keeping information from others, we can examine our motives for this secrecy. Are we withholding information out of our own self-interest, or because this is really what's best for all involved? And on a deeper level, where are we avoiding being honest with ourselves?

Move beyond polarization and the focus on "good vs. evil" and "us vs. them." Every person in the world has a heart. And each of us has a place in our heart that wants only to love and be loved. Consider the possibility that all of us are doing what we believe to be right based upon our beliefs, circumstances, and upbringing. As we focus less on blaming and judging "them" and more on supporting the highest motive in all people – including members of the power elite – we increasingly come to see that we are all one human family, and that we can choose to transform our world by working together for the good of all.

And finally, come together in groups and communities to support each other in making these positive changes. In these groups and communities, we support each other in transforming fear into acceptance and love. We exchange information about all that is being hidden from us and explore ways to move beyond polarization. We also share our life purpose and intentions and encourage each other to live these as fully as we can. And we inspire each other to take responsibility for our lives and to be the best that we can be. When we gather in community to support each other in these intentions, we join in building a growing network of inspiration and empowerment around the planet.

These suggestions emphasize changing the collective consciousness of the world by transforming both our own personal lives and the world around us. Let us do all that we can to stop destructive behaviors of the global power elite. Yet let us not overly focus on changing the outside world without first having a solid internal foundation, lest we fall into blaming others, and into the polarizing "us vs. them" ways which only further divide us. By reminding ourselves that the most powerful change starts inside each one of us, we can develop more balance and strength to then work towards positive change out in the world.

As we change ourselves from the inside out, we send ripples of inspiration and healing out into the world. By joining together in groups and communities to support each other in this, those ripples become waves which are powerfully helping to build a brighter future for all of us.

Social Responsibility

In the last 20 years, there has been a great deal of talk about "human rights", but very little discussion about the individual's responsibility to society. Society cannot provide the individual with rights and freedoms if the individual is unwilling to provide the society with responsible behavior in return. The arrival of unlimited amounts of low-cost, clean energy could be extremely destabilizing if bestowed upon an uncivil society. These developments provide unparalleled opportunities for both economic disruption and greater cooperation.

A civil society is a balancing act. On one side are the individual's needs for freedom of expression and freedom of action. On the other side are society's needs for providing everyone with a similar set of opportunities. The idea that individuals are FREE to act without restraint is a myth. No such situation has ever existed nor is ever likely to exist. There are no individual freedoms that are not actually defined or allowed by the group. Likewise, there are no group restraints that are not defined or enforced by individuals upon each other. We are each, both individuals AND members of a larger society. To enjoy freedom of action within the group, we must behave within the group definition of what is acceptable. For this to work well, the group definition of what is acceptable must be sufficiently flexible so that total conformity is not required or desired.

Currently, civil society is very loosely bound together. There are very few "group agreements" that everyone voluntarily abides by. Most people believe in the idea of "law and order" but willfully ignore these well-defined standards of behavior when they think no one else is looking. While society "expects" all of the individuals to adhere to the laws currently in-force, it has never actually ASKED each individual for 100% voluntary compliance. Until each individual is actually invested with the responsibility of providing society with honesty, civility, and integrity in their personal actions, society will never be able to provide each individual with security, peace and justice as a group environment.

The people of this world are at cross-roads. Free Energy technology offers an opportunity for everyone to have an abundance of food, fresh water, and the necessities of life. This will be a dream come true for the vast majority of the population. Just to have the opportunity to lift themselves up out of poverty, enjoy some leisure time every day, and watch the children grow up with health and safety is all they have ever wanted. But to make this come true, we all must agree to cooperate, and not to fight each other during the implementation of this transition. Unfortunately, there is a small minority of individuals on this planet who profit from keeping most people at an unreasonable disadvantage. So, there are human impediments to the implementation of the "technology of abundance."

Free Energy technology will not happen for our mutual benefit if we refuse to take personal responsibility for bringing social order into our own lives. Everyone must participate and agree. All National Governments must be constrained to live peacefully within their own borders and have benign relations with all other Nation-states. All individual, commercial, and governmental activities that prevent this from being achieved represent the real danger to society.

As social order becomes more and more unstable, the imposition of order from outside of ourselves must become more forceful. The unfortunate end result of this trend is the imposition of a total police state, whether we like it or not. When taken to its extreme, externally imposed social order is invariably experienced as tyranny by the majority of the population. And yet, the majority of the population is unwilling to impose order upon themselves, which is the only thing that would make that tyranny obsolete.

Inexpensive, abundant energy is the only thing that gives society the ability to provide for all of the needs of all of the individuals within it. But even more importantly, free energy allows individuals to provide for their own needs, so society-at-large doesn't have to. Free energy technology empowers the individual to act with strength for his or her own benefit. But this also endangers society as a whole, since these individuals may choose to act in ways that benefit themselves at society's expense. Herein lies the dilemma. Since all individuals have never even been asked to cooperate with each other, such voluntary cooperation is unlikely to emerge spontaneously. This means that the emergence of free energy technology requires the emergence of a new social contract of peaceful co-existence as well.

Free energy does NOT solve society's social problems. Most people still prefer associating and doing business with other people from their own cultural or racial group. This is an ancient preference that dates from prehistoric times. No invention will change this. Only complete inter-breeding and the blending of all races will eliminate this problem, and that will only occur thousands of years in the future. Also, many human groups have longstanding grievances with their neighbors that must be resolved before universal peace can be established on this planet. This process will take generations and an unprecedented will to cooperate and compromise before these issues can be settled to everyone's satisfaction.

So, what can we do today? What is a socially responsible course of action that strengthens social cooperation and addresses real human needs? We each must ask this question of ourselves. While dozens of changes are needed, the implementation of a "Conserve and use Natural Energy First" philosophy in our personal lives can make a real contribution to a civil society.

So, when you buy a new light bulb, don't buy the cheapest one, buy the one that uses the least energy. The next time you paint your house, don't just pick a color you like, pick a paint that reflects heat and light better to lower your energy costs. When you replace your roof, add insulation, heat reflective coatings and add skylights to interior rooms and hallways. When you buy a new car, pick one that gets 50% better gas mileage than the one it replaces. Every choice, every decision that affects your energy use pattern can reflect your socially responsible attitude.

By making these choices, we can "walk our talk" and know we are contributing to a better future. By buying these products, we can create demands in the economy that make the businesses that produce these products more successful. This trend increases innovation that makes even better products available next year. But don't wait. Buy the best products that are available right now. In a few, short years, millions of people will be changing the world with simple, everyday decisions. Be a trend setter, start now.

Today, in the United States of America, CONSUMER SPENDING is the largest force in the economy. Let's get organized and start CONSUMING the products that save energy, use less energy, and lower our energy costs today. If we lead, business and government WILL FOLLOW!

Tell your congressional representatives that you want tax incentives for small businesses that bring energy saving products to market, and that you want tax credits for buying these products. Vote for people who will implement these policy changes. Vote against people who won't.

In the last analysis, social responsibility only manifests when individuals act responsibly. Each one of us must commit our own actions to be expressed in a way that builds a better future, regardless of what anyone else may do. We cannot be responsible for other people's actions, we can only be responsible to others with our own actions.

Reference[xxix] Article - Social Responsibility. Extracted from the Internet January 18, 2019 @ http://free-energy.ws/social-responsibility/. Written by Peter Lindemann of Clear Tech, Inc. and posted on his website titled Free Energy

New Energy Resources

Summary - ***The World of Free Energy*** - Dr. Peter Lindemann
Rating–Interest 9.5 out of 10, Reliability 9 out of 10
Free

The internal combustion engine should have been obsolete many years ago! A great variety of energy devices which generate power and electricity very cheaply have been invented over the years, only to then be ruthlessly suppressed. We personally know of several genius inventors who have suffered this fate. This brilliantly written essay describes several types of new energy devices and tells you the how and why of this major cover-up. Incredibly well researched with resources listed, including patent numbers. This is another one to print and share with friends. See also Dr. Lindemann's website at http://www.free-energy.ws

---

Website - **Free Energy News** - Website Team
Rating–Interest 9.5 out of 10, Reliability 8.5 out of 10
Free

This website provides a wealth of news articles, scholarly papers, write-ups, and other key information compiled by an excellent team specializing in the field of new energy. Among the many highly informative resources you will find on this website are a listing of the top 100 new energy technologies (https://peswiki.com/congress:top-100-technologies----rd) as determined by the New Energy Congress, an excellent new energy newsletter (choose to receive informative emails daily or weekly), a wiki catalog of an amazing variety of fascinating new energy technologies and inventions, intriguing writings by hundreds of scientists and engineers who are experimenting with breakthrough energy sources, and lots more. An informative sister website at (www.newenergymovement.org) is also well worth exploring. Join the thousands of educated, pioneering souls at the forefront of the new energy revolution.

Available at http://www.freeenergynews.com

Contributing Authors

Many Individuals Who Have Made Key Contributions to this book

Kristina Borjesson - Emmy award-winning journalist and media activist

Fred Burks - Presidential interpreter, website manager, cover-up researcher

Norma Carr-Ruffino, PhD - Professor of management, author of nine books

Steven M. Greer, MD - ER physician, author, founder of Disclosure Project

Leonard Horowitz, DMD - Author, speaker, authority in public health education

Ronald Hudkins – Creator, Author, 20 year veteran, law enforcement, investigations, researcher

Michael Levine - 25-year veteran of DEA, award-winning author, radio show host

Peter Lindemann, DSc - Author, researcher, expert in field of new energy

Pamela J. Monday, PhD - Professional counselor, lecturer, researcher

Peter Phillips, PhD - Professor specializing in media, director of Project Censored

Carol Rutz - Author, lecturer, researcher, recovered mind control survivor

Peter Dale Scott, PhD - UC Berkeley professor, author, cover-up researcher

Paul Thompson - Author, researcher, creator and manager of the 9/11 Timeline

Charles S. Viar - Expert in intelligence and counterintelligence, 30 years' experience

Author Bibliographies

Kristina Borjesson – As an investigative reporter and independent producer for more than twenty years, Kristina Borjesson has worked at a variety of top media agencies including CBS network, where she won an Emmy and a Murrow Award for her investigative reporting on "CBS Reports: Legacy of Shame" with Dan Rather and Randall Pinkston. She has worked for CBS, CNN, and PBS covering a rich variety of topics including the TWA flight 800 disaster, Haiti, Fidel Castro, and crocodiles.

Kristina compiled and edited the landmark book Into the Buzzsaw—a collaborative effort of 20 award-winning journalists. Each of these brave journalists suffered from concerted corporate and/or government efforts to kill their highly revealing stories and their careers. This courageous book is by far the best, most reliable source we've seen that exposes the cover-up of some of the most important news ever reported. For an excellent two-page summary, see www.WantToKnow.info/mediacorruption.

Fred Burks – Before resigning from the US State Department in 2004 due to excessive secrecy demands, Fred Burks served many years as a language interpreter for presidents and other dignitaries. He interpreted for G.W. Bush, Clinton, Gore, Cheney, and many other top officials of the US and other countries. Having participated in secret meetings where the only people allowed were the principals and their interpreters, he has acquired important inside information and contacts.

After receiving a wealth of eye-opening information on major cover-ups from respected friends and colleagues, Fred was inspired to develop this website in October 2002. Continually opening to divine guidance, Fred has dedicated himself both to getting the word out about the cover-ups, and to inspiring others to help transform ourselves and our world through love and empowerment. He is currently executive director of the PEERS network of websites and manager of WantToKnow.info.

Fred has also helped develop and foster several inspiring communities and websites, including Moment of Love and the free Personal Growth Courses. To read some of his inspirational writings, click here. For a fun, inspiring 12-minute talk on his interpreting experiences, click here. For Fred's incredible personal journey, click here. Fred's deepest commitment in life can be summed up with these simple words: "I give all that I am into the service of All That Is."

Norma Carr-Ruffino - PhD (http://online.sfsu.edu/~ncr) – For the past 30 years Prof. Carr-Ruffino has devoted more and more time and energy to reconnecting with her soul and spirit—and to sharing her insights in seminars and books. In recent years, she's learned about major cover-ups of which few are aware, and has had a growing need to know just what our government has been doing in our name. Her pattern these days is "find the truth no matter how disturbing."

Here is a brief summary of Norma's history:

MBA and Ph.D., University of North Texas, 1969 and 1973

Professor of Management, San Francisco State University since 1973

Founder or co-founder of 3 businesses

Author of 9 books:

Making Diversity Work, 2004.

Managing Diversity: People Connections in a Diverse Workplace, 2003.

The Innovative Woman: Hot Skills for the New Economy, 2001

The Creative Intelligence Model: Building Innovative Skills, 2001

Diversity Success Strategies, 1999

The Promotable Woman: 10 Essential Skills for the New Millennium, 1997

Mujer de Empresa, 1999 (Spanish version of The Promotable Woman)

Business Students Guide, 1991 (a course supplementary text)

Writing Short Business Reports, 1980

Steven M. Greer, MD – A lifetime member of Alpha Omega Alpha, the nation's most prestigious medical honor society, Dr. Greer is the Founder and Director of the Disclosure Project. He is also an emergency physician and former chairman of the Department of Emergency Medicine at a hospital in North Carolina.

On May 9, 2001, as Director of The Disclosure Project, Dr. Greer presided over The Disclosure Project Press Conference, held at the highly esteemed National Press Club in Washington, DC. At this landmark event, over twenty military, government, intelligence, and corporate witnesses gave compelling testimony of their personal experiences regarding the existence of extra-terrestrial life forms visiting the planet.

Many millions around the globe watched or heard about this press conference through webcast and subsequent media coverage on BBC, CNN worldwide, Voice of America, Pravda, Chinese media, and media outlets throughout Latin America. Dr. Greer has also been seen and heard by millions world-wide on shows such as The Larry King Show, CBS, BBC, NTV in Japan.

Leonard Horowitz - DMD, MA, MPH – An internationally-known authority in public health education, Dr. Horowitz is the award-winning author of numerous books in the fields of emerging diseases, bioterrorism, toxic warfare, and natural healing methods. He received his doctorate from Tufts University School of Dental Medicine and a Master of Public Health degree in behavioral science from Harvard University. He currently serves as President of Tetrahedron, a non-profit educational corporation.

Dr. Horowitz's book, Death in the Air: Globalism, Terrorism and Toxic Warfare, strikes at the heart of an inbred culture of secrecy responsible for the latest forms of biochemical terrorism, mind control warfare, and ecogenocide. The revelations contained in this monumental book offer more than a reality check concerning the war against terrorism. The advice Dr. Horowitz provides gives humanity, you, and your loved ones a choice for survival. This revelatory information is available through www.tetrahedron.org. See his official website at www.DrLenHorowitz.com.

Ronald Hudkins - Is an eclectic writer. His diverse writing prowess has been influenced by his many experiences. He was drafted into the military in 1970 and remained there until 1993 when he retired honorably from the U.S. Army, Military Police Corps.

During his service and even afterwards he pursued several courses of study at several universities, including Kent State University, Maryland University, Central Texas College (European Branch), Blair Junior College, Hagerstown Junior College and Phoenix University. Despite many General Studies, he declared two majors during his schooling which included the areas of Business Administration and a Bachelor of Science in Information Technology.

In all these scholastic studies though, the true treasures he took away were not the certificates (though those are very important), but instead the experiences he had, the people he met, the foods he ate and even the places he stayed in.

In truth he is a citizen of the world and this has greatly influenced his writing. He has authored over 75 books that are listed on his author platform at http://ronaldhudkins.com/index.html

Michael Levine – Michael Levine is a 25-year veteran of the DEA (Drug Enforcement Agency) turned best-selling author and journalist. Currently, he hosts the popular Expert Witness radio show on WBAI in New York, while working on his next book. His articles and interviews on the drug war have been published in numerous national newspapers and magazines, including the New York Times, Los Angeles Times, USA Today, and Esquire.

Mike is the author of two best-selling books on his drug war experiences: the New York Times best-seller, Deep Cover and the national best-seller, The Big White Lie. He has served as a consultant and on-air expert for various national television programs in both Spanish and English, including 60 Minutes, Crossfire, MacNeil/Lehrer NewsHour, Good Morning America, and Contrapunto. Read about his riveting experiences with the DEA at www.WantToKnow.info/massmedia#levine.

Peter A. Lindemann - DSc – Dr. Lindemann became interested in new energy in 1973, when he was introduced to the work of Edwin Gray. By 1981, he had developed his own new energy systems based on variable reluctance generators and pulsed motor designs. During the 1980's, he worked off and on with both Bruce DePalma and Eric Dollard. In 1988, he joined the board of directors at Borderland Sciences Research Foundation, and served until 1999.

Currently, Dr. Lindemann is a research associate of Dr. Robert Adams in New Zealand, as well as a close collaborator with Trevor James Constable in the U.S. Dr. Lindemann is one of the foremost authorities on the practical applications of ether technology and cold electricity. His inspiring summary of the new energy field may be the best out there. See Dr. Lindemann's new energy website at www.free-energy.ws.

Robert E. Mc Elwaine - B.S., Physics and Astronomy, UW-EC, 2nd Initiate in Eckankar. Self-described Bachelor of Science in physics who wrote a series of ranting fringe science essays characteristically peppered with capitalized words for emphasis. Each essay covered topics such as alien influence on violence, free energy, coming UFO landings, and cancer cures, often carrying a message that there existed a conspiracy to suppress the information. The essays often concluded with the signature "UN-altered REPRODUCTION and DISSEMINATION of this IMPORTANT Information is ENCOURAGED, ESPECIALLY to COMPUTER BULLETIN BOARDS." McElwaine's writings stopped appearing on Usenet after 1998, although he continued writing essays up to 2003. He died at age 59 in his home in Eau Claire, Wisconsin, on 12 February 2008.

Pamela J. Monday - PhD A Licensed Marriage and Family Therapist and a Licensed Professional Counselor, Dr. Monday received her PhD in Counseling Psychology from the University of Texas. She teaches and supervises licensed therapists, and presents at conferences around the US. She serves on the boards of two non-profit organizations, Imago Relationships International, and PEERS.

Dr. Monday specializes in helping individuals, couples, and families to achieve satisfying and fulfilling lives and relationships. She empowers people to reach their potential, and to find meaning, purpose, and happiness in life. Dr. Monday has worked with numerous clients with DID (Dissociative Identity Disorder) and with other victims of physical and psychological trauma. She has worked extensively with other professionals in understanding and deprogramming survivors of little-known government mind control programs which actively induced DID. Learn more about Dr. Monday on her website at www.relationshiprepairshop.com.

Peter Phillips - PhD – A professor and Department Chair of Sociology at Sonoma State University, Prof. Phillips teaches classes in Media Censorship, Sociology of Power, Political Sociology, and Sociology of Media. He was also director of the highly esteemed Project Censored, a research group composed of 200 faculty, students, and community experts. Project Censored covers top news stories that were either ignored or downplayed by the mainstream media each year. Many editions of Censored: Media Democracy in Action were published under his skilled leadership.

Prof. Phillips regularly writes op-ed pieces for independent media nationwide and speaks on media censorship and various socio-political issues on radio and TV talk shows heard and seen across the US. He is also the national and international news editor of the North Bay Progressive in Santa Rosa, California — a monthly publication serving a five-county area north of San Francisco. Prof. Phillips' doctoral dissertation was entitled A Relative Advantage: Sociology of the San Francisco Bohemian Club.

Carol Rutz – This is the courageous author of A Nation Betrayed. This well-written, landmark book exposes a major cover-up of government mind control programs from one who knows the topic all too well. Though she bases key parts of this revealing book on her research of over 18,000 pages of declassified government mind control documents, Carol speaks from personal experience as a survivor of these secret programs.

A Nation Betrayed tells the true story of secret Cold War experiments performed on children and other unsuspecting citizens. Through both extensive research and networking with other survivors, Carol has pieced together this little-known story of the most heinous secret experiments conducted by top doctors, scientists, and government officials under the guise of national security. Carol remains active in the survivor community where she has published numerous articles and has been featured at various conferences. You can read an eye-opening 10-page summary of her harrowing encounters at www.WantToKnow.info/nationbetrayed10pg.

Peter Dale Scott – PhD, a former Canadian diplomat and English Professor at the University of California, Berkeley, Prof. Scott is a poet, writer, and researcher. He was a co-founder of the Coalition on Political Assassinations (COPA), and of the Peace and Conflict Studies Program at UC Berkeley. He has authored numerous articles and books on US foreign affairs.

His poetry has dealt with both his experience and his research, the latter of which has centered on US covert operations, their impact on democracy at home and abroad, and their relations to the John F. Kennedy assassination and the global drug trade. The poet-critic Robert Hass has written (Agni, 31/32, p. 335) that his "Coming to Jakarta is the most important political poem to appear in the English language in a very long time." Prof. Scott has won numerous awards for both his poetry and his political commentary. Learn more about him at this link.

Paul Thompson – Creator and primary author of the information-packed 9/11 timeline. His book The Terror Timeline: Year by Year, Day by Day, Minute by Minute: A Comprehensive Chronicle of the Road to 9/11–and America's Response is used as a powerful reference by many key researchers of the 9/11 movement.

The name is an alias because he wants to maintain some privacy and anonymity. He was born in California and graduated from Stanford University in the early '90s. Never previously involved in politics or activism, he began creating the timeline in mid-2002 after slowly realizing the many holes in the official story. For an excellent two-page summary of the timeline, see www.WantToKnow.info/911/9-11-timeline.

Charles S. Viar – A top expert in the fields of intelligence and counterintelligence, Charles Viar first became involved in intelligence during his service in the United States Marine Corps Reserve (1971-1973). He has been continuously involved in intelligence or intelligence related matters ever since. He holds undergraduate degrees in history, political science, and education; a master's degree in international relations; and has published several hundred articles addressing various aspects of national security.

After working at the American Security Council and the Heritage Foundation, Mr. Viar became Executive Director and President of the Security and Intelligence Foundation. Between 1985 and 1987, he had the unique privilege of studying under the late James J. Angleton, the legendary former chief of CIA Counterintelligence. Since 1989, he has been Chairman and CEO of the Center for Intelligence Studies in Washington, DC. For Mr. Viar's gripping autobiographical story, click here. For his excellent article on the history and development of mind control as an intelligence weapon, click here.

John W. Wagner – Mr. Wagner, school teacher who for many years taught in an elementary school in Ann Arbor, Michigan, carried on a long crusade to educate the American people about the scientific accomplishments of Nikola Tesla. His 3rd and 4th grade students helped with his research, and in writing letters about it. As a result, they were able to donate statues of Tesla to many Universities, such as Harvard, Yale, Princeton, MIT and others.

Acknowledgements

You and I extend heart-felt gratitude to all of the dedicated individuals and to the countless others who are deeply committed to helping spread the empowering information to all concerned citizens of the world. Together, we are all building a brighter future for ourselves, for our children, and for our world.

Cover Designs (front and rear)

courtesy of https://www.fiverr.com/pro_design190

Article References

[i] Article – The World of Free Energy By Peter Lindemann, D.Sc.. Extracted from the Internet @ https://www.wanttoknow.info/freeenergy on January 18, 2019

[ii] Article - The U.S. Government Has a Secret System for Stalling PatentsWritten By Alyssa Bereznak, National Correspondent, Technology, Yahoo Tech•December 3, 2014 Extracted from the Internet January 2, 2019 at https://finance.yahoo.com/news/the-u-s-government-has-a-secret-system-for-104249688314.html

[iii] Article - Action to Declassify Secret Energy Patents written by Gary Vesperman. Extracted from the Internet January 18, 2019 @ http://www.spiritofmaat.com/archive/feb2/prns/vesprman.htm. Gary Vesperman is a tireless writer and activist with knowledge in a vast area of subjects of interest and concern to modern technology and our way of life. His mailing address is 3123 Trueno Road, Henderson, NV 89014-3142, phone 702-435-7947. He can be reached via email at vman@skylink.net.

[iv] Article – Book – Into the Buzzsaw: Leading Journalists Expose the Myth of a Free Press – Edited by Christina Borjesson, Extracted from the Internet January 13, 2019 @ https://www.wanttoknow.info/massmedia

[v] Article - Clearing the Black Smoke of Fake News, October 28, 2018October 28, 2018. Written by Andy Lee Roth whom is the associate director of Project Censored and co-editor of the Project's newest book, Censored 2019: Fighting the Fake News Invasion. Extracted from the Internet January 13, 2019 @ https://www.projectcensored.org/clearing-the-black-smoke-of-fake-news/

[vi] Article - Media Deception & Public Perception, Transformation Index lesson plan one extracted from the Internet January 14, 2019 at https://www.transformationteam.net/ttcourse/ttcourse02a

[vii] Article - FreeTheTech Finds 5,784 Inventions Suppressed by U.S. Government Secrecy Orders, written By Sam Davis whom is editor-in-chief of Power Electronics Technology magazine and website. Posted September 14, 2018. Extracted from the Internet January 10, 2019 @ https://www.powerelectronics.com/news/sorry-primary-kilogram-we-are-dumping-you . For more information, visit the project's main website at www.FreeTheTech.org.

[viii] Article - Are You Overwhelmed or Upset With All The Craziness?. Posted on website WantToKnow.info which is an educational website. Extracted from the Internet by permission January 12, 2019 @ https://www.wanttoknow.info/overwhelmed

[ix] Free energy suppression conspiracy theory Extracted from the Internet January 15, 2019 @ https://en.wikipedia.org/wiki/Free_energy_suppression_conspiracy_theory#cite_note-16

[x] Article - Free Energy Technology Written by Robert E. Mc Elwaine, B.S., Physics, UW-EC Extracted from the Internet January 8, 2019@ https://www.bibliotecapleyades.net/ciencia/secret_projects/project119.htm

[xi] Article - A Cheap, Inexhaustible, Self-Sustaining and Non Polluting Source of Energy, written by Richard Walters, People Magazine / Energy/New Ideas section: Subtitle: A

promising new alternative energy source, neglected in the U.S. advances in the Far East. Extracted from the Internet January 8, 2019 @ https://www.bibliotecapleyades.net/ciencia/secret_projects/project014.htm

[xii] Article – The Wireless Power Grid: More Than A 100 Years In The Making. Written and posted By Julianne Geiger - Nov 19, 2017, 5:00 PM CST. Geiger is s a veteran editor, writer and researcher for US-based Divergente LLC consulting firm, and a member of the Creative Professionals Networking Group. Extracted from the Internet @ https://oilprice.com/Energy/Energy-General/The-Wireless-Power-Grid-More-Than-A-100-Years-In-The-Making.html

[xiii] New Energy Sources and Inventions. This is an extract of the book titled New Energy Sources thatwas written by Peter Lindemann, D. Sc. and summarized and edited by Fred Burks https://www.wanttoknow.info/newenergysources

[xiv] Article - Cheap power: An overnight revolution. Written by Mark Gibbs whom is an author, journalist, and contributing editor to NetworkWorld from IDG. This article posted October 2011 at https://www.forbes.com/sites/markgibbs/2013/05/20/finally-independent-testing-of-rossis-e-cat-cold-fusion-device-maybe-the-world-will-change-after-all/#69b93c3924dd

[xv] Article - The coming era of unlimited, free and clean energy. Written by Vivek Wadhwa whom is a Distinguished Fellow and professor at Carnegie Mellon University Engineering at Silicon Valley and a director of research at Center for Entrepreneurship and Research Commercialization at Duke. His past appointments include Stanford Law School, the University of California, Berkeley, Harvard Law School, and Emory University. Extracted from the Internet January 13, 2019 @ https://www.washingtonpost.com/news/innovations/wp/2014/09/19/the-coming-era-of-unlimited-and-free-clean-energy/?utm_term=.3b25b38d1d7d

[xvi] Article - Wireless Power Is Coming, PC Magazine, Written By Ajay Kumar, June 8, 2018 8:00AM EST. Extracted from the Internet January 7, 2019 from https://www.pcmag.com/article/361628/wireless-power-is-coming

[xvii] Article - Introduction to Wireless Power (WPT) Extracted from the Internet January 7, 2019 @ https://www.lifewire.com/introduction-to-wireless-power-electricity-817377 Written by Bradley Mitchell An MIT graduate who brings years of technical experience to articles on SEO, computers, and wireless networking. Updated June 01, 2018. Bradley is a former writer for Lifewire

[xviii] Article - The Great Energy Transition: The Years Just Ahead. Published September 26, 2017 by Peakoil.com News. Extracted from the Internet January 17, 2019 @ https://peakoil.com/alternative-energy/the-great-energy-transition-the-years-just-ahead

[xix] Article - Randell L. Mills – A Living Legend, Greater Than Einstein and Tesla Combined Article posted in the Libertarian News. Written by Michael Suede July 12, 2016. Extracted from the Internet January 17, 2019 @ https://www.libertariannews.org/2016/07/12/randell-l-mills-a-living-legend-greater-than-einstein-and-tesla-combined/

[xx] Article - Futurist Dr. Randell Mills Talks SunCell, Off-Grid Power, And The Future Of Job Creation. Written by Rebekah Iliff, contributing writer to the HUFFPOST and is currently the Chief Strategy Officer for AirPR, a PR technology company. Posted June 01, 2017. Extracted

from the Internet January 17, 2019 @ https://www.huffingtonpost.com/entry/futurist-dr-randell-mills-talks-suncell-off-grid_us_592ec431e4b07c4c73138706

[xxi] Article - Finally! Independent Testing Of Rossi's E-Cat Cold Fusion Device: Maybe The World Will Change After All Written by Mark Gibbs 2013-05-20 for Forbes and Posted: 2013-05-28 10:02:12. Extracted from the Internet 2 January 2019 from https://www.forbes.com/sites/markgibbs/2013/05/20/finally-independent-testing-of-rossis-e-cat-cold-fusion-device-maybe-the-world-will-change-after-all/#2b39775b24dd

[xxii] Article - Signs of Hope, by Bob Herbert for The New York Times November 23, 2009. Extracted from the Internet January 15, 2019 from https://www.nytimes.com/2009/11/24/opinion/24herbert.html

[xxiii] Article - Scalar Energy - A Completely New World Is Possible, By Bill Morgan wmorgan@nycap.rr.com. Posted 7-22-3. Extracted from the Internet January 6,2019 @ https://rense.com/general39/scalarenergy.htm

[xxiv] Article - Turning physics on its ear. Written by Tyler Hamilton, Staff Reporter for Star Business Journal. Posted Mon., Feb. 4, 2008. Extracted from the Internet January 13, 2019 @ https://www.thestar.com/business/2008/02/04/turning_physics_on_its_ear.html

[xxv] Article - Nikola Tesla - Forgotten American Scientist. Extracted from the Internet http://www.ntesla.org/ntesla/NT-P.13.html January 18, 2019. Posted in near complete entirety in honor of both the publisher John W. Wagner, school teacher and his commemorative efforts of inventor Nikola Tesla

[xxvi] Article - Nikola Tesla's 5 Lost Inventions That Threatened The Global Elite. Published January 2, 2017 by Collective Evolution. Extracted from the Internet on January 9, 2019 @ https://www.collective-evolution.com/2017/01/02/nikola-teslas-5-lost-inventions-that-threatened-the-global-elite/

[xxvii] Article – Tesla and EMF Has multiple references [51] See Tesla's FBI files at the FBI FOIA site located at https://vault.fbi.gov/nikola-tesla.
[52] "Tesla: Life and Legacy, Missing Papers," archived at PBS: http://www.pbs.org/tesla/ll/ll_mispapers.html.
[53] Tesla: Master of Lightning."" PBS documentary Dec. 12th, 2000.
[54] Box #8 of Declassified CIA documents pertaining to MKULTRA contains the following fragment: The Application of Tesla's Technology in Today's World. Obtain, online, through the National Security Archives at http://www.gwu.edu/~nsarchiv/

[xxviii] Article - Nikola Tesla Magnifying Transmitter, New York Times - March 27, 1904 Extracted from the Internet January 16, 2019 @ https://www.wanttoknow.info/nikolateslamagnifyingtransmitter

[xxix] Article - Social Responsibility. Extracted from the Internet January 18, 2019 @ http://free-energy.ws/social-responsibility/ . Written by Peter Lindemann of Clear Tech, Inc. and posted on his website titled Free Energy

Ronald Hudkins.com

Hudkins Publishing

Phoenix, AZ

Printed in Dunstable, United Kingdom

64653647R00159